Prairies, Forests, and Wetlands

A Bur Oak Original

PRAIRIES,

The Restoration of

FORESTS,

Natural Landscape

AND

Communities in Iowa

WETLANDS

BY JANETTE R. THOMPSON

Foreword by John A. Pearson

University of Iowa Press Iowa City

University of Iowa Press, Iowa City 52242

Copyright © 1992 by the University
of Iowa Press

All rights reserved

Printed in the United States of America

Design by Karen Copp

Printed on acid-free paper

96 95 94 93 92 C 10 9 8 7 6 5 4 3 2 1

96 95 94 93 92 P 10 9 8 7 6 5 4 3 2 1

Library of Congress Cataloging-in-
Publication Data

Thompson, Janette R.
    Prairies, forests, and wetlands: the
restoration of natural landscape
communities in Iowa/by Janette R.
Thompson; foreword by John A.
Pearson.
        p.   cm.—(A Bur oak original)
    Includes bibliographical references and
    index.
    ISBN 0-87745-372-1 (cloth),
    ISBN 0-87745-371-3 (pbk.)
    1. Restoration ecology—Iowa.
2. Natural history—Iowa.
I. Title. II. Series.
QH105.I8T46 1992
333.73′153′09777—dc20   92-9457
                                 CIP

Contents

Foreword by John A. Pearson : vii

Acknowledgments : ix

1. Iowa's Natural Landscape Communities : 1

2. Prairie Restoration in Iowa : 7

3. Forest Restoration in Iowa : 42

4. Wetland Restoration in Iowa : 78

Epilogue : 103

Appendix 1. Prairie Seed and Plant Sources : 105

Appendix 2. Tree Nurseries and Seed Dealers in Iowa : 107

Appendix 3. Wetland Seed and Plant Sources : 113

Appendix 4. District Foresters : 114

Bibliography : 115

Index : 131

Foreword

BY JOHN A. PEARSON

IOWA IS THE "beautiful land," once covered by a continuous mosaic of prairies, savannas, forests, and wetlands and now famous worldwide for its bountiful harvest of crops and livestock. This success has not been without cost: natural areas in the heartland have been pushed almost entirely off of the biggest, flattest, most fertile lands and generally persist today only where it is too steep, too rocky, too dry, too wet, or too isolated to plant crops, graze livestock, construct roads, or build houses. Because deep, fertile, gently rolling soils are so prevalent in our state, this means that remaining natural areas—with some notable exceptions such as the Loess Hills and the more rugged parts of extreme northeast Iowa—tend to be small and far apart.

One of the consequences of this situation is that plants and animals which depend on natural areas face an uncertain future due to isolation as well as simple lack of habitat. Lacking wings or long legs, small mammals, amphibians, and reptiles might be unable to colonize an empty patch of suitable habitat simply because they "can't get there from here" across broad expanses of inhospitable territory. Among plants, some species such as orchids may be represented by only one or a few individuals within a single small patch, meaning that opportunities for pollination must rely upon the proximity of other patches. Without neighbors, small, isolated populations become what biologist Daniel Janzen has called "the living dead," slated to disappear when the current generation eventually dies off.

Conservation of natural areas and their associated biological diversity depends on two fundamental tools: keeping the natural areas that remain and replacing at least some of those that have been lost. Keeping natural areas involves a spectrum of activities generally termed "protection and management," including purchase of land by conservation agencies and wise stewardship of natural lands in both public and private sectors, all focused on maintaining or improving existing natural areas. Replacing natural areas means putting a prairie, a forest, or a wetland where it is presently nonexistent, such as a vacant lot, an abandoned field, or a drained basin . . . it is what Janette Thompson calls "reconstruction."

This book is about reconstructing the elements of Iowa's natural landscape. Reconstruction can contribute to solutions to the problems of isolation and habitat shortage by creating new patches to serve as permanent homes for plants and animals and also by placing "stepping stones" and corridors between widely separated natural patches. The style of reconstruction promoted by this book is as important as its content: Thompson emphasizes the need for *authenticity* of reconstructed communities by matching them with their natural soils, climatic conditions, and species diversity. In her discussion of prairies, for example, she points out that the ideal goal of a reconstruction is to restore not just a few species of dominant grasses but dozens or even hundreds of species of native plants, including a rich array of native forbs (broad-leaved wildflowers); the plantations of big bluestem and Indiangrass that we casually call "restored prairie" are but distant approximations of this ideal. Significantly, she also discusses the need to utilize local genetic resources as plant materials for reconstructions. As efforts such as the Iowa Ecotype Project—a joint project by the Soil Conservation Service and the Roadside Management Program based at the University of Northern Iowa to gather and mass-produce stocks of native Iowa prairie plants—progress, these goals will become less utopian and more affordable.

This book is a grand synthesis of the ecological and horticultural knowledge about natural communities in Iowa which was previously scattered in scientific journals, in informative yet obscure agronomic handbooks and pamphlets, and in the heads of numerous botanists, ecologists, wildlife biologists, foresters, and other experts. Janette Thompson is to be commended for her patience and perseverance in assembling a useful and coherent summary from diverse and perhaps even contradictory sources. (Anyone who has attempted to reconcile the opinions of two or more experts can surely appreciate this challenge.) Her thoughtful discussion of the ecological backgrounds and prospects for local reconstruction of Iowa natural communities will be widely appreciated by serious naturalists and backyard restorationists alike.

Acknowledgments

THE IDEA OF CREATING a how-to manual for landscape restoration was that of an interdisciplinary team of scientists who were part of the Research Unit on Landscape Ecology (RULE) at Iowa State University. RULE committee members included Arnold van der Valk, Dick Schultz, Don Farrar, Paul Wray, Gary Hightshoe, Ken Lane (all from Iowa State), Steve Lekwa (Story County Conservation Board), Richard Bishop (Iowa Department of Natural Resources), and Trelen Wilson (formerly Story County Conservation Board). Thanks are due to these persons because, in a sense, this book is their brainchild. I would also like to acknowledge partial financial support from the Research Unit on Landscape Ecology during the initial stages of manuscript preparation. Continued financial support (Department of Forestry, Iowa State University) and encouragement from Dick Schultz carried this project from early drafts to its final form.

Although I remain responsible for any errors or oversights herein, I would like to recognize the efforts of several persons who informally reviewed portions of the manuscript for me: Susan Galatowitsch, Beth Middleton, Wayne Fisher, Paul Christiansen, Don Farrar, Steve Lekwa, Richard Pope, Richard Schultz, Paul Wray, and Jim Dinsmore. Thanks are also due to formal reviewers Diana Horton, Lon Drake, Richard Baker, John Pearson, and Cynthia Hildebrand, who suggested numerous improvements to the text, and to Rosaura Gonzalez for help with some of the illustrations.

Prairies, Forests, and Wetlands

1 : Iowa's Natural Landscape Communities

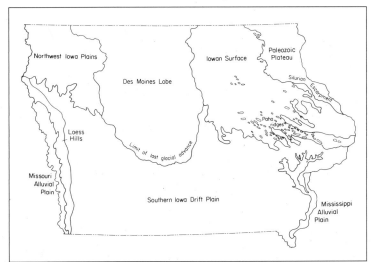

Figure 1.
Landforms of Iowa.
By Jean C. Prior,
courtesy Iowa
Department of
Natural Resources.

THE ARRIVAL OF Euro-American settlers in the mid-1800s signaled the beginning of a period of rapid change for the landscape ecology of Iowa. Prior to this, native vegetation communities in Iowa represented a natural evolutionary response to the complex of factors that define the environment—geology, landforms and soils, climate, other organisms, and time. Early observers indicated that the central Iowa landscape was dominated by prairies and prairie-wetland complexes, with forest communities adjacent to water-courses throughout much of the state and forests or savannas dominating the landscape in parts of the southern and eastern portions of the state. Through time, the conversion of more than 90 percent of the original landscape to agricultural and urban uses has nearly eliminated these natural communities (Farrar 1981). New understanding of the functional significance of these ecosystems has contributed to interest in reconstructing, restoring, and maintaining areas of native vegetation communities.

This book is intended to be a nontechnical natural history and a guide for amateur restorationists interested in the reconstruction of prairies, forests, or wetlands in the state of Iowa. The bibliography includes numerous publications on each of the different natural communities considered here for those who seek or require more detailed information. An effort has been made to present as many options for methods of site preparation and establishment of native vegetation as practical in a book of this size. Readers should recognize, however, that natural ecosystems are not simple machines, and the protocol for their reintroduction and management cannot and should not be reduced to simple recipes. In almost every case the appropriate restoration techniques will be entirely site-specific.

Discussion of plant species throughout the text is limited to vascular plants. Both common and scientific names of the species discussed are given in the text. Unless otherwise noted, the scientific nomenclature follows that of the Great Plains Flora Association (1986).

Some people reserve the term *restoration* for management or improvement of areas with poor-quality existing native communities and prefer the term *reconstruction* for reintroduction of native species to areas where the natural flora do not occur. The two terms are used interchangeably here since it is strongly suggested that these communities only be reintroduced to areas in which they previously appeared.

GEOLOGY AND LANDFORMS

The physical environment of Iowa has changed dramatically over geologic time. Much of the bedrock underlying surficial materials in Iowa is sedimentary rock, indicating the presence of vast seas for extended periods of time long in the past. More recent geologic deposits above this bedrock framework are comprised of unconsolidated materials which arrived during the Ice Age. During this time, several cycles of glaciation affected the northeast and north-central United States. The series of glacial advances and retreats from about 2.5 million to about 10,000 years before present did much to create modern Iowa landforms. Iowa was partially or entirely covered by glacial ice several times during the Pleistocene epoch. In the warm season, floods of sediment-loaded water discharged from the ice margins carved the large floodplains and valleys associated with the Missouri and Mississippi rivers. Melting slowed in the cold season, and water-carried sediments were deposited. Windstorms removed relatively fine (silt-size) materials from river floodplains and dispersed them over large areas of the Midwest, forming extensive loess deposits. Glacial, windblown, and water-laid deposits cover approximately 97 percent of Iowa's land surface (Prior et al. 1982).

The landforms of Iowa are shown in figure 1. Iowa's oldest landscape features appear in the extreme northeast corner of the state, a region referred to as the Paleozoic Plateau (Prior 1991). The plateau-like uplands and steep, narrow valleys in this area are determined by underlying sedimentary rock formations of sandstone, limestone, and dolomite (Prior et al. 1982). Patchy remnants of glacial drift occur in this area, which is covered by a layer of loess.

The undulating plains across southern (Southern Iowa Drift Plain), northwest (Northwest Iowa Plains and Tazewell Drift area), and part of northeast (Iowan Surface) Iowa are covered with glacial deposits that represent all known ice advances into the state except the most recent two (Prior 1991). The glacial deposits have been exposed to weathering for some 500,000 years, so that these regions have a well-developed, integrated system of streams which effectively drains the area. A younger covering of loess blankets much of the region. In extreme western Iowa (Loess Hills region), the loess has accumulated in unusual thickness and dominates the terrain. Much of the loess in Iowa was deposited between 28,000 and 14,000 years before present.

The most recent episode of glaciation to affect the state occurred between 14,000 and 12,000 years before present and left a tongue-shaped region of glacial till deposits in north-central Iowa called the Des Moines Lobe (Prior 1991; Prior et al. 1982). Stream erosion is just beginning to establish a drainage pattern in this region, and numerous closed, undrained depressions still exist. Several distinctly aligned bands of ridges (moraines) were deposited within and along the edges of the lobe as stagnant ice melted.

SOILS

Most soils in Iowa have formed in the geological deposits already mentioned: glacial till (an unsorted mixture of sand, silt, clay, gravel, and boulders), loess, and alluvial (water-laid) deposits on bottomlands and floodplains. A large proportion of Iowa's soils formed under the influence of prairie vegetation. Below-ground decomposition of the fibrous root systems of grasses and forbs enriched the topsoil with organic matter, forming soils with deep, dark-colored, nutrient-rich surface horizons. Soils formed under the influence of forest vegetation are less extensive in Iowa and typically have less organic matter enrichment in surface horizons and are lighter in color. Largely because of the relative youth of the geologic materials in which the soils are formed and the influence of prairie vegetation over much of the state, Iowa's soils are among the most productive in the world (Fenton and Miller 1982).

CLIMATE

Wood fragments and pollen analyses indicate major shifts in the dominant vegetation in Iowa over the last 30,000 years, which reflect changing climatic conditions (Prior et al. 1982). Spruce and pine species were common when cool, moist conditions prevailed during and following episodes of glaciation, in particular between 30,000 and 20,000 years before present, and spruce was abundant again 15,000 to 11,000 years before present (Graham and Glenn-Lewin 1982). Grasses became the dominant life form approximately 8,000 to 9,000 years ago as the climate became significantly hotter and drier and, perhaps, as fires restricted the growth of woody plants. Some climatic moderation is suggested by the reinvasion of deciduous trees, especially oak, about 3,000 years ago (Prior et al. 1982).

Iowa's current climate is one of marked seasonal variation because of its interior continental location and northern latitude. In the warm season, prevailing southerly flows of air from the Gulf of Mexico carry about 70 percent of Iowa's annual precipitation. In the cold season, masses of dry arctic air arrive from the northwest (Waite and Shaw 1982). The average annual temperature is about 49 degrees F, with an average extreme range from − 16 to 99 degrees in the central portion of the state. Iowa experiences frequent and rapid weather changes caused by the arrival of one or two different air masses per week. Average annual precipitation ranges from 25 inches in the northwest to about 35 inches in the southern and eastern portions of the state, with a statewide average of 21 inches during the growing season. The length of Iowa's growing season varies from 145 days in the north to 170 days in the southeast (Waite and Shaw 1982).

LANDSCAPE COMMUNITIES

Estimates of the size of pre-Euro-American settlement areas of prairies, forests, and wetlands in Iowa are based on observations by surveyors and naturalists active during the mid-1800s to early 1900s. Among early naturalists, Thomas Macbride, Bohumil Shimek, and Louis Pammel did much to describe the kinds and extent of vegetation in the state. Based on writings of Shimek (1911) and later of Hayden (1945), it is widely held that prairies covered close to 85 percent (just over 30 million acres) of Iowa prior to the arrival of Euro-American immigrants (Smith 1981). This estimate probably includes an area of 7.6 million acres of mixed prairie and wetlands in north-central and northwest Iowa, 2 to 3 million acres of which could have been wetlands (Bishop 1981). Additional wetland areas (open water of lakes, oxbows, backwater areas, and meandering watercourses of major rivers and tributaries) would have accounted for substantial acreage, but estimates of this area are difficult to find. One of the best available estimates concerning original forest acreage is that of Pammel (1896), who reported 5 to 6 million acres of forest coverage in Iowa shortly after Euro-American settlement. These estimates of the different vegetative types may best serve to indicate the overwhelming dominance of prairie in the state in the early 1800s. Recent inventories of prairies, wetlands, and forests in Iowa indicate that approximately 30,000 acres, 110,000 acres, and 1.9 million acres respectively of these native communities remain (Pearson 1990; Smith 1981; Bishop and van der Valk 1982; Kingsley 1990). With respect to current inventories, estimates of original cover serve as a general indication of the degree to which all three natural ecosystems have been nearly obliterated by human activity in the past 150 years.

BENEFITS OF REESTABLISHING AREAS
OF NATIVE VEGETATION

Some values of prairies, wetlands, and forests represent direct economic assets: for example, timber value, recreational value, or value of food harvested as fish and game. Other values associated with these natural communities may in fact be far greater: the genetic value of species that may represent millions of years of natural selection (e.g., adaptation to climate, disease and insect resistance), the scientific benefits derived from the study of natural ecosystems, the role of the ecosystems in providing wildlife habitat and food chain support, erosion and sedimentation control, carbon and nutrient retention, as well as the maintenance of biological diversity for future generations, often referred to as a "heritage" value. Many species which are considered endangered, threatened, or declining in Iowa can survive only in specific prairie, forest, or wetland habitats. In addition, reestablishment of native prairie, forest, and wetland communities affords an opportunity to address contemporary local concerns about agricultural diversification and concerns about environmental quality on a global scale. More specific benefits associated with restoration of each of these communities will be discussed in subsequent chapters dealing with the respective vegetation types.

2 : Prairie Restoration in Iowa

Figure 2. A mixture of late-blooming forbs and grasses on Stinson Prairie in Kossuth County.

HISTORY OF PRAIRIES

A LTHOUGH THE prairie appeared as a recognizable natural plant community probably in the Oligocene epoch of the Tertiary period (about 25 million years before present, after uplift of the Rocky Mountains roughly 65 million years ago contributed to the development of the present continental climate), pollen analyses suggest that the prairie most recently became a dominant vegetative community in Iowa only about 8,000 to 9,000 years ago (Baker and van Zant 1978; Smith 1981). Early ecologists and botanists recorded that the tallgrass or "true" prairie ecosystem extended over parts of Ohio, Indiana, Illinois, Michigan, Wisconsin, Minnesota, most of Iowa, extreme eastern North and South Dakota, eastern Nebraska, southern Manitoba and Saskatchewan, parts of Kansas, Missouri, Arkansas, and much smaller areas of Oklahoma and Texas (Transeau 1935). Iowa is the only state that lies entirely within the natural region of the tallgrass prairie (Smith 1981), and early documents indicate that approximately 85 percent (close to 30 million acres) of the state was covered by prairie vegetation at the time of Euro-

American settlement (Shimek 1911a; Hayden 1945). The predominance of the prairie plant community over much of central North America for thousands of years prior to Euro-American settlement has been attributed to climatic characteristics of the region (Transeau 1935), to widespread occurrence of fire initiated by lightning or by Native Americans (Sauer 1950), and to animal activities such as grazing by bison and other large herbivores (Edwards 1978).

Transeau (1935) introduced the concept of prairie as a climatic vegetation "climax" in the North American Midwest, suggesting that the region occupied by tallgrass prairie could be outlined within an area having a narrow range (.6 to .8) of precipitation to evaporation ratios. Other climatic variables which may contribute to the predominance of prairie species in this region are seasonal patterns of rainfall distribution, with peak precipitation in spring and autumn and little precipitation in winter and summer (Risser et al. 1981), and periodic extended (up to ten years) and/or severe droughts recurring approximately every thirty years (Duvick and Blasing 1981; McClain 1986). The warm-season prairie grasses native to the central states are well adapted to withstand such droughty conditions.

Both topography and climate on the Great Plains favored frequent and extensive fires before Euro-American occupation of the area. Nearly level to gently rolling land surfaces allowed fires to move rapidly across the landscape, with only a few widely spaced rivers to act as natural firebreaks (Risser et al. 1981; McClain 1986). Accumulation of dry grasses fueled the spread of fire in the windy weather of early spring and late fall. Settlers gave many accounts of fires begun by Native Americans early in spring to stimulate the growth of grasses that would attract large herbivores and to trap or drive animals, especially bison, late in the fall (Becic and Bragg 1978). The role of fire in fostering prairie vegetation in the Midwest is supported by the relative fire tolerance of prairie plants (some specific fire adaptations of prairie flora will be mentioned in a later discussion of prairie maintenance), the fact that small aboveground shoots of woody species common in the region are usually killed by fire and originally appeared in greatest numbers on the lee side of natural firebreaks such as rivers, and the spread of woody vegetation immediately following the suppression of fires by homesteaders in the mid-1800s (Risser et al. 1981; Smith and Christiansen 1982). Fire is now used in prairie management to help prevent invasion by woody plant species.

The prairie soils were ideal for crop production. Although turning the prairie sod was a difficult task requiring large breaking plows pulled by several yoke of oxen (Coffin 1902), the ready availability of the steel moldboard plow in the 1850s led to the rapid conversion of the prairie for agricultural purposes: nearly 30 million acres of Iowa prairie were converted in eighty years (1850–1930), for an average of 375,000 acres per year (Smith 1981). Interest in the origin, distribution, and restoration of Iowa's prairies was stimulated by the writings of Bohumil Shimek in the early 1900s (e.g., Shimek 1911a, 1911b, 1925) and led to efforts by Crane and Olcott (1933), Hayden (1945, 1946), Anderson (1945), Aikman (1949), and others to preserve tracts of nearly undisturbed prairies and to restore degraded prairie areas. In 1946, the state of Iowa purchased its first prairie preserve, Hayden prairie, now a 240-acre tract located in Howard County. A number of small (less than 200 acres) prairie tracts acquired by the state, county conservation boards, and the Iowa chapter of the Nature Conservancy since the mid-1940s constitute the bulk of Iowa's state prairie preserves, for a total of some 5,000 acres (Pearson 1990). Other small unprotected prairie remnants persist along railroad rights-of-way, in areas cut off from cultivated ground by roads or railroads, in old cemeteries and prairie hayfields, on steep slopes in stream-dissected landscapes (especially in the Loess Hills of western Iowa), or as fringes around undrained pothole wetlands (Schennum 1986). The best available estimate indicates that in all, probably only about 30,000 acres of Iowa's original 30-million-acre tallgrass prairie remain (Pearson 1990).

Efforts toward prairie restoration are not new to the state of Iowa. Considerable small-scale prairie restoration work was undertaken in the 1930s, 1940s, and 1950s (e.g., Anderson 1936, 1945; Landers and Christiansen n.d.). Much of the current work on restoration of native grasses (and in some cases the more complete prairie plant community, including forbs as well) is being conducted by county conservation boards and roadside management programs (e.g., Iowa Department of Natural Resources 1988).

TYPES OF PRAIRIES IN IOWA

The prairie is a complex ecosystem dominated by grasses and forbs (forbs are perennial wildflowers, usually broad-leaved plants) but also includes some species of shrubs and sedges (Nichols and Entine

1976). Prairie plant community ground cover averages about 60 percent grasses, 35 percent forbs, and 5 percent shrubs (Ahrenhoerster and Wilson 1981). By weight, grasses represent close to 90 percent of the annual plant growth and are also the dominant plants below the soil surface, with a dense network of fine roots that may reach depths of 6 feet or more (Smith and Christiansen 1982).

In spite of the relative dominance of the grass family, species richness is a hallmark of the prairie ecosystem. Risser et al. (1981) suggest that representative areas of the true prairie contain about 250 species of higher plants. Shimek (1931) documented close to 265 species as the bulk of Iowa's prairie flora. About 72 different species of grasses are commonly found in Iowa's prairies, with about 50 common species from the daisy (composite) family and 25 common species of the legume family (Smith and Christiansen 1982). Representatives of the rose, buttercup, milkweed, mint, sedge, and parsley families are also present in Iowa's prairies.

In addition to species diversity, other salient features of the major species of prairie vegetation are their broad ecological amplitude (tolerance to a wide range of environmental conditions) and large geographic ranges—for example, from southern Texas to southern Saskatchewan (Risser et al. 1981). These features are related both to the plasticity of prairie plants and to the ecotypic variation in their genetic makeup. In addition, 95 percent of prairie species are perennials, with life spans commonly up to twenty years and sometimes exceeding one hundred years (Ahrenhoerster and Wilson 1981; Risser et al. 1981; Costello 1969).

A complex topographic-soil moisture gradient is important in determining plant community structure and species distribution within the prairie (Crist and Glenn-Lewin 1978; White and Glenn-Lewin 1984). Accordingly, prairies have been divided into three general types (similar to the classifications of Weaver 1954 and Curtis 1959): wet prairies, mesic (moist) prairies (including wet-mesic and mesic prairies), and dry prairies (including dry-mesic, Loess Hills, gravel hills, and sand prairies) (White and Glenn-Lewin 1984). A relatively simple moisture gradient may occur locally on hillslopes that extend from relatively dry uplands to moist lowlands (fig. 3); a similar but more gradual gradient occurs from east to west across the entire tallgrass prairie region (Rock 1981).

Somewhat distinctive groups of prairie species are associated with

Figure 3. Changes in prairie species along a topographic gradient are noticeable on Anderson Prairie in Emmet County. Mesic and dry-mesic species including asters, coneflowers, and rough blazing star dominate the foreground; wet-mesic species, most visibly sunflowers, are common downslope (center).

the general prairie types. These groups of plants can sometimes be used to identify the prairie segment(s) that historically occupied a site (if any native plants remain) and are among the most appropriate plants to be reintroduced to their respective moisture segment in prairie restoration. Many other common species can grow vigorously under a wide range of soil moisture conditions and may appear in several of the prairie types. Only a few of the more typical species present in Iowa prairies (unless otherwise noted) will be mentioned in the brief discussion of the prairie types which follows; a more detailed list of prairie plants that can be used for restoration is given in table 1.

WET PRAIRIES

The moist soils of wet prairies are characterized by more available moisture than that supplied directly by precipitation—for exam-

Table 1. Selected Species Suitable for Prairie Restoration in Iowa

Species	Habitat	Establishment	Seed Conditioning
Grasses			
Bluejoint reedgrass	W (except ext. w)		
Prairie cordgrass	W, W-M	easy / early (sm. qty.)	by division
Prairie panic grass	W, W-M, M (common nc)	easy / early	
Switchgrass	M	easy / early (sm. qty.)	
Big bluestem	M	easy / early	
Prairie dropseed	M (common nc and sw)	mod. / middle (transplants)	
Indiangrass	M, D-M	easy / early	
Canada wild rye	M, D-M	easy / early	
Scribner's panic grass	M, D-M	easy / early	
Rough dropseed	M, D-M (except ext. ne)	mod. / middle	
Little bluestem	M, D-M, D	easy / early	
Purple lovegrass	M, D-M, D (except ext. nw)	easy / early	
Needlegrass	D-M	easy / early	cold-moist
June grass	D-M, D	easy / early	
Sideoats grama grass	D-M, D	easy / early	cold-moist
Forbs			
New England aster	W, W-M	easy / early (sm. qty.)	cold-moist
Bottle gentian	W, W-M (most frequent nc)	mod. / middle	cold-moist
Purple meadow rue	W-M	easy-mod. / middle	cold-moist
Prairie blazing star	W, W-M, M, D-M (rare nw, common elsewhere)	mod. / middle	cold-moist
Ragwort	W, W-M (nc and se)	mod. / middle	cold-moist

Table 1. *(continued)*

Species	Habitat	Establishment	Seed Conditioning
Black-eyed susan	W-M, M, D-M (except ext. nw)	easy/early (sm. qty.)	cold-moist
Rosinweed	W-M (mostly s half)	easy/early	cold-moist
Golden alexanders	W-M, M	mod./middle	cold-moist
Grayhead prairie coneflower	W-M, M	easy/early (sm. qty.)	cold-moist
Sawtooth sunflower	W-M, M	easy/early (very sm. qty.)	cold-moist
Wild bergamot	W-M, M	easy/early	cold-moist
Cup plant	W-M, M (common except ext. nw)	easy/early	cold-moist
Rattlesnake master	W-M, M (common except ext. w)	easy/early	cold-moist
Compass plant	W-M, M, D-M (most common w half)	easy/early	cold-moist
Sky-blue aster	W-M, M, D-M (se two-thirds)	mod./middle	cold-moist
Tall cinquefoil	M (except ext. ne)	mod./middle	cold-moist
White wild indigo	M (except rare ne and nw)	easy/early	cold-moist, scarify
New Jersey tea	M (except ext. se)	mod./middle (transplant)	cold-moist, scarify
False sunflower	M	easy/early	cold-moist
Jerusalem artichoke	M (infreq. n, common s)	easy/early (sm. qty.)	cold-moist
Illinois tick-trefoil	M, D-M (s half and e)	easy/early	cold-moist, scarify
Rough blazing star	M, D-M (common e two-thirds)	mod./middle	cold-moist
Purple coneflower	M, D-M (sc and se)	easy/early	cold-moist
Smooth blue aster	M, D-M	mod./middle (transplants	cold-moist
Stiff goldenrod	M, D-M (except ext. se)	easy/early (very sm. qty.)	cold-moist

Table 1. *(continued)*

Species	Habitat	Establishment	Seed Conditioning
Tall boneset	M, D-M (frequent s, rare n)	easy / early	cold-moist
Frost aster	M, D-M (common s half)	easy / early (sm. qty.)	cold-moist
False boneset	M, D-M	easy / early	cold-moist
Showy sunflower	M, D-M, D (common w half)	easy / early (sm. qty.)	cold-moist
Prairie phlox	M, D-M, D	mod. / middle (transplants)	cold-moist
Leadplant	M, D-M, D	mod. / middle (transplants)	cold-moist, scarify
Purple prairie clover	M, D-M, D	mod. / middle (transplants)	cold-moist, scarify
Prairie anemone	M, D-M, D (common nw half)	mod. / middle (transplants)	cold-moist
Hoary vervain	M, D-M, D	easy / early (very sm. qty.)	cold
Plains wild indigo	M, D-M, D	mod. to difficult (transplants)	cold-moist, scarify
Round-headed bush clover	D-M	easy / early	cold-moist, scarify
Prairie coreopsis	D-M (infreq. se, common elsewhere)	easy to mod. / middle	cold-moist
Whorled milkweed	D-M	easy / early (sm. qty.)	cold-moist
Old-field goldenrod	D-M, D (except ext. nw)	easy / early (sm. qty.)	cold-moist

Note: Species are listed by moisture segment going from wet to dry: W = wet, W-M = wet-mesic, M = mesic, D-M = dry mesic, D = dry, ext. = extreme. Unless otherwise noted, range includes all of Iowa. If range is limited, it is indicated as east (e), west (w), north-central (nc), etc.

Sources: Eilers n.d.; Ahrenhoerster and Wilson n.d.; Betz 1986; Schramm 1978; Farrar 1989; McClain 1986; Landers and Christiansen n.d.

Figure 4. A typical closed depression or prairie pothole on Doolittle Prairie in Story County. Wet prairie species present include prairie cordgrass and blueflag iris in the foreground and smartweed in the central portion of the pothole.

ple, wet prairies on flat, low-lying areas adjacent to rivers, ponds, and marshes may receive additional moisture from overland flow (Ahrenhoerster and Wilson 1981; Rock 1981). In the prairie pothole region of north-central Iowa (the Des Moines Lobe), small remnants of undisturbed wet prairies remain as borders around undrained pothole wetlands (Schennum 1986), with typical wet prairie plant communities grading into the wet meadow and mudflat annual plant communities of the wetlands (fig. 4). Species of the wet prairie include sedges (*Carex* spp.), New England aster (*Aster novae-angliae*), bluejoint reedgrass (*Calamagrostis canadensis*), marsh muhly grass (*Muhlenbergia racemosa*), yellow stargrass (*Hypoxis hirsuta*), cowbane (*Oxypolis rigidior*), common mountain mint (*Pycnanthemum virginianum*), prairie cordgrass (*Spartina pectinata*), purple meadow rue (*Thalictrum dasycarpum*), blueflag iris (*Iris virginica*), swamp saxifrage (*Saxifraga cernua*), and bottle gentian (*Gentiana andrewsii*) (Rock 1981; Smith and Christiansen 1982; White and Glenn-Lewin 1984) (fig. 5).

Figure 5. Typical species of wet and wet-mesic prairies, bottle gentian and sunflower, on Hayden Prairie in Dickinson County.

MESIC PRAIRIES

White and Glenn-Lewin (1984) identify four plant community types within mesic prairies, distinguishing wet-mesic and mesic plant communities in impeded and unimpeded drainage types. These prairie communities appear on lower slopes in areas with relatively high relief (slopes of more than 10 percent), in upland prairies with only slight relief or fine-textured soils, and in the highest positions in otherwise wet landscapes (fig. 6).

Several important species occur in all of the mesic prairie types: big bluestem (*Andropogon gerardii*) and prairie dropseed (*Sporobolus heterolepis*) are dominant among the grasses, although little bluestem (*Andropogon scoparius*) may also be abundant. In wet-mesic plant communities (in both impeded and unimpeded drainage types), common species may also include sedges, wild strawberry (*Fragaria virginiana*), sawtooth sunflower (*Helianthus grosseserratus*), common mountain mint, grayhead prairie coneflower (*Rati-*

Figure 6. Mesic prairie species, most notably rattlesnake master, compass plant, big bluestem, and rigid goldenrod, on Doolittle Prairie in Story County.

bida pinnata), Canada goldenrod (*Solidago canadensis*), heath aster (*Aster ericoides*), meadow bedstraw (*Galium obtusum*), and golden alexanders (*Zizia aurea*). In mesic plant communities (again, in both impeded and unimpeded drainage types), common species include Scribner's panic grass (*Dichanthelium oligosanthes* var. *scribnerianum*), leadplant (*Amorpha canescens*), sky-blue aster (*Aster oolentangiensis*), false sunflower (*Heliopsis helianthoides*), Indiangrass (*Sorghastrum nutans*), showy sunflower (*Helianthus laetiflorus*), and golden alexanders (White and Glenn-Lewin 1984).

DRY PRAIRIES

Dry prairies occur on the upper slopes of high-relief prairie landscapes or on sites with rapidly permeable soil over coarse substrates well above the water table (White and Glenn-Lewin 1984). Dry prairie plant communities are characterized by a more open canopy and plants of lower stature than wet or mesic prairie communities.

Figure 7. The canopy of a dry prairie is relatively open. Switchgrass and purple-top are common grasses on Marshall County's Marietta Sand Prairie.

White and Glenn-Lewin (1984) distinguish three types of dry prairies: dry-mesic, gravel hill, and sand prairies (fig. 7).

Dry-mesic prairie plant communities frequently occur on well-drained upper slope positions and may include little bluestem, big bluestem, needlegrass (*Stipa spartea*), prairie dropseed, Indiangrass, sideoats grama grass (*Bouteloua curtipendula*), Scribner's panic grass, leadplant, heath aster, yellow puccoon (*Lithospermum incisum*), prairie coreopsis (*Coreopsis palmata*), showy sunflower, prairie wild rose (*Rosa arkansana*), and prairie goldenrod (*Solidago missouriensis*) (White and Glenn-Lewin 1984; Glenn-Lewin et al. 1984).

Special dry prairies in Iowa occur in the Loess Hills region at the western edge of the state. The assemblage of species distinctive of Loess Hills prairies includes plains muhly grass (*Muhlenbergia cuspidata*), sideoats grama grass, sand dropseed (*Sporobolus cryptandrus*), yucca (*Yucca glauca*), great-flowered beardtongue (*Penstemon grandiflorus*), skeleton weed (*Lygodesmia juncea*), scarlet guara (*Guara coccinea*), sand lily (*Mentzelia decapetala*), red three-awn

(*Aristida purpurea*), and stiffstem flax (*Linum rigidum*) (Smith and Christiansen 1982; Mutel 1989; Eilers n.d.).

Gravel hill prairies are present on steep, gravelly upper slopes associated with the moraines on the Des Moines Lobe. Species in this community may include little bluestem, plains muhly grass, blue grama grass (*Bouteloua gracilis*), hairy grama grass (*Bouteloua hirsuta*), needlegrass, leadplant, prairie goldenrod, stiff goldenrod (*Solidago rigida*), old-field goldenrod (*Solidago nemoralis*), fleabane (*Conyza canadensis*), aromatic aster (*Aster oblongifolius*), and blazing star (*Liatris punctata*) (White and Glenn-Lewin 1984).

True sand prairie flora appears on eolian sand ridges associated with major rivers in the state. The surface soil is very dry, but abundant moisture may be present at 6- to 9-feet depths beneath the surface. Sand prairie communities may include little bluestem, erect dayflower (*Commelina erecta*), fameflower (*Talinum rugospermum*), purple milkwort (*Polygala polygama*), pink milkwort (*Polygala incarnata*), knotgrass (*Paspalum setaceum*), sand dropseed, rough blazing star (*Liatris aspera*), and evening primrose (*Oenothera biennis*) (White and Glenn-Lewin 1984; Glenn-Lewin et al. 1984; Eilers n.d.).

BENEFITS OF PRAIRIE RESTORATION

The benefits accruing from prairie restoration have been reiterated many times since first suggested by early naturalists such as Shimek and Hayden. As Hayden (1945) pointed out, the native prairie was a naturally self-perpetuating plant community which helped to build and maintain the fertility and ideal physical properties of some of the world's most productive soils. This observation points to two features which are compelling reasons for the restoration of prairies: maintaining the reservoir of genetic information in plants and animals well adapted to their environment and allowing opportunities for study of the interactions between plant communities and soil properties in an ecosystem with sustained productivity characterized by simultaneous soil improvement. Particularly important for roadside managers, the prairie plant community is one which naturally subdues weeds so that less effort (and less herbicide application) is required for weed control (fig. 8). Other benefits of prairie restoration include the value of such natural areas to wildlife, natu-

Figure 8. A roadside prairie in northern Boone County, composed primarily of big bluestem, Canada wild rye, Indiangrass, and prairie cordgrass.

ral heritage value and use of prairies as interpretive areas, and the aesthetic values associated with this plant community.

THE GENETIC RESOURCE

Iowa's native prairie flora alone includes close to 400 species (although the more common species probably number in the range of 200), and the natural prairie ecosystem is composed of hundreds of additional species of insects, birds, and mammals (Shimek 1931; Wagner 1975; Platt 1975; Howe 1984). These species have co-evolved over thousands of years of natural selection and represent genetic systems inherently well suited to local climatic conditions, with built-in resistance to pests and diseases. The combination of biological diversity and genetic adaptation makes the prairie a reservoir of genetic information with enormous potential use (Wagner 1975).

SUSTAINED PRODUCTIVITY AND SOIL IMPROVEMENT

The prairie plant community not only protects against erosion but enriches the chemical and physical properties of the soil (e.g., promotion of aggregate formation and stability of structure; see Jastrow 1987). Earlier, Hayden (1945) suggested that one valuable function of native prairie plant communities would be for use as an experimental check of the fertility and structure of soils that have undergone cultivation for many years. So far, the data indicate that for large-scale and long-term production, humans have been unable to mimic the sustained productivity of the native prairie with maintenance of soil properties, much less soil improvement. As Lebovitz (1987) has noted, "The prairie runs on sunshine and accumulates soil; the local fields are dependent on fossil fuels and are losing soil."

Contemporary workers propose that the prairie community serve as a model for diversified systems of agricultural production (the "perennial polyculture" concept; see, for example, Drake 1978; Lebovitz 1987; Piper 1988) which are less demanding of human-supplied (usually fossil fuel) energy than traditional crops and farming systems and much less damaging to the environment. Others have suggested that long-term maintenance of soil fertility and soil physical properties might best be achieved by rotating crops with prairie fallow periods (Nelson 1986). Use of prairie plant communities to ameliorate severely degraded soils, such as in strip-mine reclamation and road construction cut/fill areas, has also been suggested (Schramm and Kalvin 1978; Dancer 1985).

Thus, the restoration of prairies can afford additional opportunities for the study of natural processes within the prairie ecosystem as well as providing a benchmark to evaluate the effects of long-term cultivation of soils. In an era of increasing concern about both economic diversification and sustainability of agricultural practices, the scientific benefits of prairies provide ample justification for additional prairie restoration in Iowa.

WEED CONTROL

Once established, prairies can dominate a site (in the absence of major disturbances other than fire), preventing invasion by many weeds (exceptions include very aggressive alien weeds such as smooth brome and leafy spurge). A combination of factors—including the species diversity of the prairie community, the low fertility requirements of prairie species (compared with exotic species), and the fact that many prairie plants are perennials and are naturally "genetically engineered" for local conditions—contribute to the value of native prairie communities for weed control. In an effort to reduce the economic and environmental costs of roadside mowing and herbicide application, restoration of prairie plant communities (or simply native grasses) in ditches as a means of weed control has met with considerable success (e.g., the Iowa Integrated Roadside Management Program housed with the County Assistance Center at the University of Northern Iowa). In most cases, after establishment of prairie vegetation, periodic prescribed burns or occasional mowings are the only requirement to maintain nearly weed-free prairie roadsides.

WILDLIFE HABITAT

Even though there is some small-scale habitat diversity because of different canopy layers in the prairie, generally speaking, the structural simplicity of grasslands does not offer the degree of niche diversification possible in more complex habitats such as forests. Still, prairies can provide important resources for animals that occupy broad ecological niches. Of 102 mammalian species native to prairies, only 18 utilize strictly grassland habitats, 39 utilize ecologically diverse habitats, and 45 utilize a combination of grassland and forest habitats (principally prairie-forest border areas) (Risser et al. 1981). Relatively few birds are uniquely associated with the true prairie; only about 20 species are regularly recorded as resident breeders and nesters (Risser et al. 1981; Braband 1986). Several mammals and birds indigenous to prairies, with niches largely restricted to prairies, have been extirpated from Iowa as a result of habitat destruction: bison, elk, greater prairie-chicken, long-billed curlew, and the marbled godwit are among these species. A number of strictly grassland nesting birds are threatened or have declining populations, for example, short-eared owl, Henslow's sparrow, loggerhead shrike, and upland sandpiper. Other species naturally associated with prairies which still appear in Iowa seem to have adapted well to nonnative habitats: red fox, least weasel, opossum, northern harrier, bobolink, eastern and western meadowlarks, dickcissel, sedge wren, and grasshopper sparrow are examples (Risser et al. 1981; Howe 1984). Additional species which have been observed recently in Iowa prairie preserves include badger, prairie skink, thirteen-lined ground squirrel, grasshopper mouse, meadow jumping mouse, white-tailed jackrabbit, masked shrew, Franklin's ground squirrel, western harvest mouse, meadow vole, prairie vole, long-tailed weasel, plains pocket gopher, striped skunk, spotted skunk, raccoon, plains garter snake, American toad, leopard frog, American kestrel, ring-necked pheasant (although not a native species), snow bunting (in winter), eastern kingbird, brown-headed cowbird, common yellowthroat, vesper sparrow, and song sparrow (Platt 1975; Braband 1986).

Reports on the invertebrate fauna of prairies indicate a much more diverse group of organisms, with conservative estimates for above-ground insects alone on the order of 3,000 species (Risser et al. 1981). As prairie areas were destroyed in Iowa, populations of some insect species also declined, among the most noticeable perhaps are now-rare butterflies such as skippers that depend on specific prairie plants for egg-laying (Howe 1984).

Additional prairie restoration in Iowa could provide much-needed areas of permanent vegetation and might encourage reestablishment of viable populations of species which have been extirpated (see, for example, Wooley 1984 on reintroduction of the greater prairie-chicken). To support viable populations of relatively large species, it may be necessary to reconstruct very large areas of prairie, such as the proposed Walnut Creek National Wildlife Refuge.

SOCIAL VALUES

Other potential benefits from establishment of prairie areas include use as interpretive areas and outdoor biological laboratories (Graham 1975) and maintenance of biological diversity for future generations. The aesthetic value associated with the native prairie plant community (forbs which are in bloom seven months of the year and grasses which are beautiful year-round) has been sufficient impetus for many small prairies to be established in urban settings as relatively low-maintenance landscape areas. The natural attractiveness of prairies has also been considered in promoting tourism in the state.

PRAIRIE ESTABLISHMENT

Although it is not feasible to restore the vast expanse of prairie that once covered Iowa, it is possible to preserve the genetic resource of native prairie plants, restore the natural fertility of prairie soils, and reestablish habitat for grassland-dwelling wildlife on a smaller scale. An ideal restored prairie community in the botanical sense alone would require an area close to 200 acres (it would require closer to 1,000 acres to restore the full fire-bison interactions of the original tallgrass prairie) and contain at least 200 plant species from propagules obtained within a 50-mile radius of the restoration site (the importance of local genotypes will be emphasized in a later discussion of sources of propagules). The cost and availability of land and local plant materials are usually the most limiting factors in efforts to simulate the biology and functions of the true prairie. The amount of time required to restore a prairie depends on characteristics of the site (e.g., weed populations present prior to the restoration effort) as well as on the desired species diversity of the prairie (establishment of "late-seral" prairie plants may take ten years or longer according to Betz 1986 and Schramm 1978). Factors to consider in planning a prairie restoration project include: site selection (suitability of soil), species to be planted, site preparation (mechanical, chemical, and combination methods for seedbed preparation), planting methods, sources of propagules, and maintenance of the prairie after it is established.

SITE SELECTION

Prairie restoration is most appropriate on sites formerly occupied by prairie; given that about 85 percent of Iowa was covered by prairies at the time of Euro-American settlement, there is no lack of suitable sites for restoration.

As previously discussed, soils formed under the influence of prairie vegetation had or have special characteristics which may but do not always persist after many years of cultivation. Prairie-influenced soils typically had deep (greater than 18 inches), dark, nutrient-rich surface horizons and are often classified in the order Mollisols in the USDA system of soil classification (USDA 1975). County soil surveys contain soil maps, descriptions of soil map units, and other information necessary to identify potential prairie restoration sites based on soil characteristics. Due to intensive cultivation, accelerated loss of topsoil may have changed the surface soil characteristics of former prairie soils (Fenton 1983), particularly in steeply sloping landscape positions; in any case, consultation with local Soil Conservation Service personnel may help in evaluating site suitability.

Even in areas where it appears that most prairie vegetation has been destroyed, it is sometimes possible to find prairie plants in

Figure 9. Round-headed bush clover is a distinctive component of this railroad right-of-way prairie in Boone County.

ditches and along railroad rights-of-way (fig. 9) and fencerows (Betz 1986); in addition to soil characteristics, such plants can serve as indicators for prairies and perhaps for the type (moisture segment) of prairie that would be most appropriate for a particular site.

Other factors to consider in site selection are adjacent land use, the amount of available sunlight, the size of the potential restoration site, and especially location with respect to natural firebreaks. Consideration of the possibility of pesticide drift or runoff from neighboring properties is especially relevant in Iowa. It may be necessary to establish some kind of buffer or to plant a concentration of very hardy prairie species on sites bordering areas of intensive agricultural activity. Shade from neighboring trees may also interfere with establishment of prairies. Prairie plants thrive on sunshine, and sites receiving 75 percent or more of each day's available sunlight have the greatest potential for successful prairie establishment (Ahrenhoerster and Wilson 1981).

Although small plantings of prairie species are attractive as land-

Figure 10. An "urban prairie" near the College of Design on the Iowa State University campus. Sunflowers, sideoats, coneflowers, blazing star, and some exotic species are included.

scaping and garden plants (fig. 10), restoration of a marginally functional prairie ecosystem probably requires at least 10 to 20 acres, and much larger sites are most desirable for establishment of different prairie communities and creation of wildlife habitat (Levenson 1981; McClain 1986). The larger the area, the more authentic and complex the prairie can be. A number of persons active in prairie restoration recommend dividing a large site into several years of plantings (Landers and Christiansen n.d.; Ahrenhoerster and Wilson 1981; the "installment plan" of Schramm 1978) because of the effort required for site preparation and initial maintenance, year-to-year variations in weed seed germination, year-to-year differences in availability, variety, and viability of prairie seed, and the benefit gained from experience along the way. An additional recommendation to facilitate prairie maintenance is to select a site with natural firebreaks (a road, stream, plowed field, or closely mowed lawn) on at least two sides of the proposed prairie area (Ahrenhoerster and Wilson 1981).

McClain (1986) cautions those attempting prairie restoration on recently cropped fields to be aware of the possibility of herbicide carryover, which can damage prairie plant seeds and seedlings. Such fields should be allowed to lie fallow for at least one year.

SPECIES SELECTION

Species diversity on the restored prairie will depend on the size of the area and range of habitat types represented on the site, species native to the site, the availability of propagules (seeds or seedlings), and the cost of propagules (which can be prohibitive, especially for some of the forbs). For restoration of the true prairie, the biological optimum species diversity is the maximum diversity possible using local genotypes. The minimum species diversity required for a functional prairie is unknown. Plant species should be carefully matched to soil moisture conditions. Some prairie species are fairly aggressive and can withstand initial intense competition with weeds; these species can be used in the early stages of prairie restoration. A partial list of species which may be appropriate for prairie restoration in Iowa and their respective habitat types, natural range in the state, ease of establishment from seed, and seed conditioning requirements is given in table 1. More complete lists of prairie species (especially forbs) can be found in Rock (1981), Risser et al. (1981), and Schramm (1978).

SITE PREPARATION

Site preparation is usually the most labor-intensive phase of prairie restoration. Existing vegetation should be carefully evaluated, particularly on former pastures that may not have been plowed. On these sites, it may be worthwhile to conduct a burn (see discussion of prescribed burning) and wait one season to see what species appear.

For relatively fast establishment of prairie plants from seed on sites that have been intensively cultivated, best results have been obtained with a firm, weed-free seedbed (Landers and Christiansen n.d.; Rock 1981; Lekwa 1984; Betz 1986). Elimination of weeds and weed seeds is by far the most difficult aspect of site preparation. Several methods of seedbed preparation have been used successfully, including mechanical, chemical, and combinations of mechanical and chemical techniques. It is often necessary to begin seedbed preparation in the fall prior to sowing seed in spring, so in general, planning for a restoration project should begin at least one year before the first planting (Schramm 1978). Some success has also been reported for prairie establishment by transplanting seedlings, which can be accomplished with less site preparation (Ripp 1985; Nuzzo 1978); however, this method of planting is itself very labor-intensive and probably is not practical for large-scale prairie restoration. A more detailed discussion about using seedlings as opposed to seeds is reserved for a later section on planting methods.

MECHANICAL METHODS. The mechanical methods described here assume spring sowing of prairie seed (generally considered the most successful), but they can be modified to accommodate fall planting. For small-scale restoration (less than one-quarter acre), plots should be rototilled and kept weed-free for one growing season prior to planting seed (Lekwa 1984). Newly germinating weeds can be eliminated by periodic tillage. For larger areas, seedbed preparation may depend on previous land use. If the site was used the previous year for row crops, shallow spring disking two or three times (the last time just before planting prairie seeds) is recommended (Landers and Christiansen n.d.). On former pastures or areas with a mixture of annual and perennial weeds, it is necessary to kill the existing sod completely before attempting to seed in prairie species. One possible approach involves late-summer mowing to a height of about 12 inches, followed by fall plowing to a depth of at least 8 inches and shallow spring disking (at two- to three-week intervals) up to the time of planting (Landers and Christiansen n.d.; Lekwa 1984; Betz 1986). Betz (1986) and Rock (1981) obtained good results when spring disking was followed by harrowing (to level the area), using a roller or cultipacker to firm the seedbed just prior to and just after planting seed. These approaches have the disadvantage of leaving bare soil exposed, with the risk of losing the seed in heavy rain, and can only be recommended without reservation on nearly level sites.

CHEMICAL AND COMBINATION METHODS. No-till chemical (herbicide) seedbed preparation is possible for both small- and large-scale prairie restoration. Application of herbicides should be done with utmost caution by certified individuals using materials strictly according to label specifications. Landers and Christiansen (n.d.) and Lekwa (1984) report initial success in prairie plant establishment (especially grasses) using glyphosate (Roundup at 3 quarts/acre) to kill existing vegetation a week to eighteen days before planting prairie seed. The seed was then drilled directly into the soil, and the standing dead plant material was left in place. Lekwa (1984) suggests subsequently mowing the dead vegetation to provide a light mulch. This technique is particularly well suited for restoration on long or steep slopes, as the standing dead vegetation provides some protection against erosion in the interval between elimination of existing vegetation and the establishment of prairie seedlings.

Several combinations of tillage and herbicide application have been used in restoration projects. Again, these methods can be adapted for use before spring or fall planting. By spraying herbicides, then tilling (shallow tilling, leaving a firm seedbed), and reapplying herbicides after new weeds germinate, more weeds can be eliminated before prairie seeds are planted (the "spray-disk-spray" method, Ahrenhoerster and Wilson 1981). Simpler variations of this method, "spray-disk" and "disk-spray," have also been used. Woehler and Martin (1978) report successful establishment of prairie grasses and forbs from seed by applying herbicide (glyphosate) early in May, followed by plowing, disking every ten to twelve days, harrowing, and then planting prairie seed approximately three weeks after the initial herbicide application. Cox (1987) plowed, disked, and then applied herbicide (glyphosate) shortly before planting prairie seed. Again, these approaches have the disadvantage of leaving bare soil exposed after seeding.

PLANTING METHODS

Many of the prairie species listed in table 1 are relatively easy to establish from seed (figs. 11, 12). Use of seedling transplants is usually only recommended for very small-scale restoration projects (less than one-quarter acre) or for adding very sensitive high-quality species (forbs) that are difficult to establish in the earliest stages of prairie restoration (Schramm 1978).

SEEDING TECHNIQUES. For restoration of prairie areas using seed, consideration must be given to seed treatment prior to planting, the timing of the planting, the amount or rate of seeding, the method of sowing seed, and whether to sow an annual cover crop at the same time.

It is possible to hand-collect local seed for use in prairie reconstruction or to collect seed on a larger scale with equipment such as the Grin Reaper. Depending on the collection site, it may be necessary to obtain legal permission to collect seed, and a general rule of

Figure 11. One-year-old prairie plants from seed (black-eyed susan, false boneset, stiff goldenrod, and other species) on a well-tended research plot at the Field Extension Education Laboratory (I.S.U. Extension) in Boone County.

thumb is not to take more than one-fifth to one-third of the seed crop for common species in a given year and to take only a few individual seeds, if any, from rare plants.

Seed obtained from a commercial source has usually been pretreated by one or more of the following processes: stratification (cold storage at about 34 degrees for six to sixteen weeks), scarification (abrasion of the seed coat by chemical or mechanical means), and inoculation (introduction of nitrogen-fixing bacteria to prairie legume seeds). If seed is collected from local prairie remnants (only where permissible!), proper conditioning is essential to achieve adequate germination in the first year (Schramm 1978). Seed-conditioning treatments are given for most species listed in table 1. The majority of prairie grasses and forbs will germinate the first season after

Figure 12. A first-year restoration project from seed broadcast over a disked area, untended. Black-eyed susans are among the few forbs that flowered; more species will "appear" in subsequent years. F.E.E.L. plots in Boone County.

planting when they have been dry or moist stratified (seed mixed with sand or vermiculite, moistened for damp stratification) and stored in plastic containers at 32 to 41 degrees for at least six weeks (Rock 1981; Schramm 1978). Betz (1986) and Woehler and Martin (1978) report adequate germination of many species using bulk seed mixtures stored field-moist over the winter in rodent-proof containers in an unheated building. Seed which is moist stratified should be checked frequently to detect and eliminate mold or fungus (Lekwa 1990).

If selective seed conditioning is possible, scarification and inoculation of legume seed just before planting will enhance germination and early growth of seedlings. Scarification can be accomplished relatively simply by one of two methods: (1) wrapping the seed in cheesecloth and submerging it in water that has just been boiled (do not continue to boil the water after the seed has been added), then allowing the water to cool to room temperature before removing the seed, or (2) abrading the seed coat by swirling the seed in a large covered can with a sheet of fine sandpaper glued on the inside or by anchoring a sheet of fine sandpaper on a working surface and gently sanding the seed with a second sheet of fine sandpaper on a sanding block for about fifteen seconds (McClain 1986; Christiansen and Landers 1966).

After scarification, legume seeds may be inoculated with cultures of nitrogen-fixing bacteria. Some legumes (e.g., leadplant, wild indigo, scurf pea) require fairly specific inocula, while others (e.g., round-headed bush clover, white and purple prairie clover) can be treated with wide-spectrum inocula. Wide-spectrum inocula can usually be purchased from seed dealers, but the more specific inocula may be difficult to find. Generally, legume seeds can be inoculated by mixing them with water, a sugar solution, or milk at a rate of 1.5 teaspoons liquid for each 2.2 pounds of seed and then mixing thoroughly with the inoculant (which comes in powder form). Using milk or sugar has the disadvantage of making the seeds more attractive to rodents. For small quantities of seed, this process can easily be done in a wide-mouthed jar (McClain 1986). Seed should be planted as soon as possible after inoculation.

TIME OF PLANTING. The most success in germinating and establishing of prairie plants from seed in Iowa has occurred when the seed was sown in late May to mid-June (Landers and Christiansen n.d.; Lekwa 1984; Ahrenhoerster and Wilson 1981), although successful seedings have been done as late as early July. Weeds (cool-season, early germinating species) can be reduced if the site is disked up to the time of planting. Warm summer days provide optimum conditions for the growth of most prairie plants because they are then better able to compete with weedy species. It is also possible to sow prairie seed late in the fall (after November 1) or earlier in the spring, but poorer results can be expected due to seed predation over winter for fall plantings and greater competition from early germinating weeds in both cases (Landers and Christiansen n.d.; Ahrenhoerster and Wilson 1981).

SEEDING RATE. Seeding rates for prairie restoration can be calculated on a bulk seed basis or, if clean seed is obtained from a commercial source, on a pure live seed basis. A very wide range of recommended seeding rates appears in the literature; generally, heavier seeding rates are recommended for more rapid establishment, for higher-quality sites, for sites with relatively fine-textured soils, and/ or if seed is broadcast rather than drilled. Grass to forb seed mix ratios ranging from 3 : 2 to 1 : 2 have been recommended (Liegel and Lyon 1986; Rock 1981); in any particular project the ratio used will probably be determined by the availability and cost of forb seed. Care should be taken not to seed too heavily with the more aggressive species of grasses (and some forbs as indicated in table 1) since this could result in the elimination of some of the more competition-sensitive plants as the prairie develops (Rock 1981).

For small areas, seeding rates have been given as numbers of pure live seeds per square yard. Planting density recommendations vary from about 100 pure live seeds (PLS) per square yard to about 600 PLS per square yard (these rates are for mixed grass and forb seeds) (Landers and Christiansen n.d.; Lekwa 1984; Woehler and Martin 1978). For larger areas, rates from 14 to 18 pounds PLS per acre have been used, with 3.5 to 15 pounds PLS per acre of grass seed and the remainder forb seed (Cox 1987; Bragg 1978; Ahrenhoerster and Wilson 1981; Rock 1981). Recommendations for the use of bulk seed vary from 20 to 41 pounds per acre (Ahrenhoerster and Wilson 1981; Betz 1986).

For general purposes, Rock (1981) suggests the following bulk seed proportions: 5 to 10 pounds per acre of mixed grass seed (as many species as appropriate to the site) and 10 to 20 pounds per acre of mixed forb seed (an ideal that will probably be difficult to obtain). On sites that may be susceptible to erosion during the establishment phase, the addition of an annual cover crop such as wild rye (at 1 to 2 pounds per acre) is also recommended (Rock 1981).

Under ideal conditions (e.g., adequate soil moisture), germination of many prairie plants can be expected in ten days to two weeks. Only the more easily established forbs (as indicated in table 1) germinate fairly rapidly in such mixed plantings; others may take years to germinate and even longer to flower.

SEED-PLANTING METHODS. Prairie seed can be hand-broadcast, disseminated by small rotary spreaders or other mechanical devices that will scatter the seed evenly over the site, or drilled (Schramm 1978; McClain 1986). The seed must be dry and relatively clean for good dispersal by mechanical devices. If the seed is not drilled into the ground, it is necessary to "set" the seed using a harrow and cultipacker or other roller to lightly cover the seeds with soil (McClain 1986; Betz 1986; Rock 1981). Mixing the seed with sand or other coarse material (e.g., with a 2 : 1 sand to seed ratio) can help ensure spreading the seed evenly if it is broadcast on the site (Rock 1981; Ahrenhoerster and Wilson 1981).

Several seed drills are designed for use with the light fluffy seeds of prairie plants: the Nesbit, Rangeland, Truax, Great Plains, and Marliss drills are commercial units that have been used successfully with prairie seed (McClain 1986; Iowa Department of Natural Resources 1988). Use of a seed drill is usually recommended for large-scale restoration efforts. Seed drills are available for loan or rent in several Iowa counties—conservation commission district biologists or local Soil Conservation Service personnel can probably advise on the availability of this equipment (their offices are usually listed in telephone directories under County Government).

Drilled prairies can have an undesirable row appearance (Betz 1986), but in most cases this becomes less noticeable after two or three years (Schramm 1978). To avoid the row effect, some workers have drilled grass seed and then broadcast forb seed over the same site (e.g., Woehler and Martin 1978); others have used a large salt-spreader device mounted on an all-terrain vehicle to broadcast grass and forb seed efficiently over relatively large areas (10 to 60 acres) (Betz 1986). For use with broadcast seeders, prairie seed may need to be "debearded" (have the seed awn removed) to flow through the seeder properly; debearded seed is available for some species of grasses from commercial seed suppliers at an increased cost.

TRANSPLANTING TECHNIQUES. Transplanting is very labor-intensive and is not recommended as a general-purpose method for restoration of large prairie areas. Use of this technique is warranted for introducing very competition-sensitive species in later stages of prairie restoration or for species enrichment of degraded prairie remnants or recently seeded areas (Rock 1981; Schramm 1978; Landers and Christiansen n.d.). Production of seedlings for transplanting on a reasonable scale may require use of greenhouse facilities or raised-bed nurseries (fig. 13). For greenhouse culture, stratified seeds should be planted in a 50:50 sand to potting mix substrate (Nuzzo 1978) in pots or flats in fall or very early in spring (late February–early April) to be outplanted in May or June (Rock 1981). According to McClain (1986), the seedlings should have four or five true leaves and a well-developed root system prior to transplanting. One- to two-year-old seedlings and/or rootstocks are also available commercially for some species (see subsequent section on sources of propagules).

Figure 13. Large-scale production of native prairie species using raised beds at the Illinois Department of Conservation's Mason State Tree Nursery in Topeka, Illinois. Prairie seed can be harvested from the plants, or they can be used as transplant stock. Photo by R. C. Schultz.

Similar to results with seeding techniques, the success of transplanted seedlings will depend on the amount of competition from weeds at the outplanting site (Anderson 1945; Christiansen and Landers 1969). Seedlings have become well established when transplanted into clearings that were either sprayed with herbicides or cultivated prior to the introduction of prairie plants (Schramm 1978; Nuzzo 1978). Ripp (1985) reports successful establishment of several species after handplanting rootstocks immediately following a prescribed burn. Transplanting just after a burn on an established prairie is recommended since the dense thatch associated with prairie grasses can make planting difficult (McClain 1986). Another suggestion to enhance survival of seedlings to be transplanted into areas where prairie plants are already present is careful placement of transplanted material in the appropriate soil moisture segment of the prairie at some distance from clumps of more aggressive grasses and forbs (McClain 1986). Transplanted seedlings need to be watered regularly and may need to be protected by fencing until they become established (Schramm 1978; Ahrenhoerster and Wilson 1981).

SOURCES OF PROPAGULES

Philosophically, there are important reasons for using locally collected propagules (usually seed): restoration of a site to its original condition implies reestablishment of what was originally present, the local genotype is well adapted to the site and is capable of producing vigorous stands, introduction of commercially selected varieties may contaminate the genetic makeup of local populations by hybridization, and particularly aggressive horticultural varieties may jeopardize the existence of native plants and result in monotypic stands of selected strains (Schramm 1978; Nuzzo 1981; Ahrenhoerster and Wilson 1981). Many restoration workers are concerned about community integrity and have noted that the introduction of alien genotypes can lead to unnatural succession patterns, changes in food web structure, or concomitant introduction of disease or other pathogens into the prairie community (e.g., Nuzzo 1981). Ahrenhoerster and Wilson (n.d.) recommend finding a seed source within 50 miles of the restoration site. This ideal is difficult to realize for any but very small-scale prairie establishment projects where local hand-collection of seed is feasible. This approach should be taken, however, for any prairie restoration work in the vicinity of a high-quality prairie preserve or remnant. In fact, for planting in these sites, the best solution may be to get permission to collect conservative amounts of seed from the remnant itself. For reconstructions far removed from existing prairies, Schramm (1978) suggests seeking propagules within an area 100 to 200 miles north or south and 150 miles east or west of the restoration site, being careful of large climatic gradients within these distances. In this framework, it is increasingly possible to find commercial sources of large enough quantities of seed for sizable restoration projects. However, it will still be important to avoid using horticultural varieties because in general these have not been selected for use in mixed species plantings (Lekwa 1990).

There are a number of seed dealers in and near Iowa who sell prairie seed. A partial list of sources appears in appendix 1. Most of these firms sell native grasses; a smaller number also carry forb (wildflower) seed (this is indicated on the list). Some of these sources may also provide small numbers of seedlings, but supplies can vary from year to year. Again, try to avoid horticulturally selected strains when purchasing seed, plant only those species that are native to the restoration area, and make an effort to obtain seed of geographic origin near the planting site that has been collected in an ethical manner.

PRAIRIE MAINTENANCE

Initial maintenance, primarily for weed control during the first two growing seasons, can speed prairie establishment. In most cases, even with some kind of initial weed control, a minimum of three growing seasons after sowing seed is needed before a recognizable prairie plant community can be expected (Woehler and Martin 1978). The majority of prairie species will not flower until at least the third year if planted as seed (Landers and Christiansen n.d.). After the prairie is established, it will probably require little maintenance other than periodic burning to control invasion by woody plants and to help to control weeds.

On very small sites, hand weeding during the first year or two is usually adequate. On larger sites, some workers recommend mowing with a flail-type mower set to cut weeds above the height of the prairie plants (6 to 12 inches) or haying (mowing and removing the mulch) in late June or early July the first year and again in May and/or June the second year (Landers and Christiansen n.d.; Lekwa 1984). Care should be taken to avoid windrowing the cuttings, especially if using a rotary mower (Lekwa 1990). These techniques are suggested because early growth (especially the first year) of perennial prairie species is predominantly underground; since prairie species do not develop tall shoots until the second or third growing season, it is possible to cut annual weeds back without damaging the young plants. Although mowing is not absolutely necessary, it is effective in controlling weeds such as lamb's-quarter and velvetleaf and may hasten prairie establishment by limiting competition for light (Schramm 1978).

PRESCRIBED BURNING

As discussed earlier, the midwestern prairie evolved under the influence of frequent fires, and prairie species can compete with exotic perennial species and woody plants under management regimes that include regularly scheduled controlled fires. Although fire does not discriminate between the native and exotic plants on a site (all aboveground plant parts are burned), prairie species recover more rapidly than woody plants.

Burning has several beneficial effects (besides helping to control weeds and woody plants) for prairie species, primarily the removal of accumulated dead plant material. Removal of the dead mulch associated with prairie plants leads to earlier warming of the soil surface in spring and exposes soil for germination of new seedlings (Landers and Christiansen n.d.; Lekwa 1984; Hulbert 1986). Periodic fires can actually stimulate the flowering and productivity of many prairie species, especially grasses (Richards and Landers 1973; Hill and Platt 1975; Hulbert 1986). Continually burning in the early spring, however, can have deleterious effects on cool-season and early blooming prairie species, so an occasional fall or late spring burning should probably be part of the regimen. Although the optimum interval between burns to enhance species diversity, productivity, and flowering is unknown (Hulbert 1986), it is generally recommended to plan for a burn once every three to five years after the prairie is well established (Lekwa 1984). Bragg (1978) proposes a three-year cycle, burning approximately one-third of the prairie area each year in order to leave undisturbed areas for wildlife. The optimum frequency and timing of burning will be site-specific, particularly with respect to which weed species are a problem on the site (e.g., late spring burning is more effective in controlling bromegrass, often a problem in Iowa prairie remnants). The consensus is that spring—sometime in late March or during most of April—usually provides optimum conditions for burning: the ground is cool and moist, many prairie plants are still dormant, weather conditions are more reliably favorable, few birds have begun nesting, reptiles and amphibians are still hibernating, and good responses by prairie species are usually observed (Bragg 1978; Schramm 1978). As suggested earlier, however, some variation in this routine may be necessary to preserve cool-season prairie plants. In areas where burning is not possible because of proximity to dwellings or other structures, haying can serve as a substitute. In order to most closely simulate some of the important effects of fire (e.g., warming of the soil surface in spring), removal of the cuttings by raking and baling is recommended (Diboll 1986; Lekwa 1984).

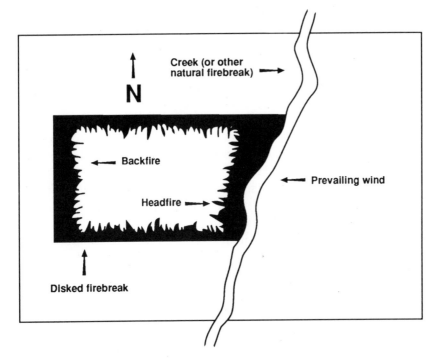

Figure 14. Proper placement of firebreaks, headfire, and backfire for a controlled burn. Modified from McClain 1986.

PROTOCOL FOR CONTROLLED BURNS

Proper planning is absolutely necessary to conduct a safe prairie fire. According to McClain (1986) and Hulbert (1978), preparation for a burn should include the following.

1. Construction of firebreaks where natural firebreaks are not present. Roads, ditches, ponds, and creeks can serve as "natural" firebreaks (fig. 14). Constructed firebreaks can be grassy areas mowed periodically during the growing season or disked, plowed, or raked to provide 6- to 8-foot strips surrounding the burn area. In areas particularly susceptible to erosion, it is possible to create fireguards without tillage (see Hulbert 1978). Setting backfires (fires that burn into the prevailing wind) along constructed firebreaks is mandatory to widen the firebreaks before starting the headfire (the main fire that burns with the prevailing wind).

2. Provision for adequate crew and equipment. The number of per-

sons required to contain the fire will depend on the size of the proposed burn and the amount of accumulated fuel in the area. Christiansen (1990) recommends using a crew of no fewer than four persons—one to set the main fire, two to control the fire, and one to monitor the backfire. Use of hand tools, including fireswatters and backpack pumps, is usually adequate. Provide a source of extra water for the pumps. Crew members should be healthy (exclude persons with heart conditions, respiratory disease, or high blood pressure), wear natural fibers rather than synthetics (which can melt against the skin when near a source of radiant heat), and be thoroughly oriented to their duties. Keys should be left in any vehicles near the fire so that the vehicles can be moved quickly in case of an emergency. Two-way communication with all crew members should be possible while conducting the burn.

3. Investigation of the local weather conditions. Conduct a burn only when weather conditions are appropriate. Wind velocity should be less than 15 miles per hour (but strong enough that the wind direction is stable), the relative humidity greater than 40 percent, and the temperature less than 70 degrees. For any sizable fire, a weather radio for continuous monitoring of changing weather conditions is a necessity.

4. Coordination with local officials such as the fire department and/or county sheriff's department. It may be necessary to obtain an open burning permit from local agencies. If persons on the crew have never participated in a controlled burn, be sure to seek assistance from an experienced person before the first burn takes place (consult local fire departments or county conservation board or Department of Natural Resources biologists). More detailed information on prairie burning is given in Pauly (1982).

COST OF PRAIRIE RESTORATION

The cost of restoring a prairie can vary dramatically depending on the desired species diversity and, if planting seed, whether seed is purchased or hand-collected, the seeding technique used, and the rate at which seed is sown. In most cases, both forb and grass seed are priced by species (for smaller projects it is possible to get packets of mixed seed). Current (1991) prices for forb seed are as high as $200 to $400 per pound of pure live seed (although often forb seed is only available in bulk at prices ranging from about $40 to $250 per pound). Prices for grass seed are much lower, averaging about $11 per pound of pure live seed. The cost of both grass and forb seed can vary substantially from year to year depending on the availability of seed and the level of demand. Additional costs of $35 to $130 per acre per year (equivalent of land rent) can be expected if retiring the land from other uses.

Site preparation costs (mechanical or chemical) for restoration projects larger than about 5 acres were on the order of $125 per acre in 1991. Establishment costs (seed and planting activities) averaged close to $500 per acre but again are extremely variable depending on the price of seed. All told, establishing a prairie from seed may cost as much as $625 per acre. Hand-planting transplant stock currently costs close to $700 per acre, with an additional cost of $2 to $3 per plant for stock. Prairie maintenance costs may be minimal, averaging about $12 to $18 per acre (for a three-year burning rotation) after the establishment period (Reed and Schwarzmeier 1978; Widstrand 1985; Twarok 1989; Kaduce 1990).

3 : Forest Restoration in Iowa

Figure 15. Oak-cedar forest on a bluff overlooking the Mississippi Riv bottomland forest, Allamakee County

HISTORY OF FORESTS

CLIMATIC CHANGES (primarily those associated with episodes of glacial activity) and relative migration rates for different species have largely determined which tree species were historically dominant in Iowa and those which are presently common. Pollen analyses and wood fragments recovered from well cores indicate that spruce and pine trees dominated forests in Iowa under cool and moist conditions during and following advances of glaciers between 30,000 and 20,000 years before present and spruce and larch again between 15,000 and 11,000 years before present (Prior et al. 1982; Graham and Glenn-Lewin 1982). By 9,000 years before present, species such as oak, elm, maple, and basswood were more common than conifers. Grassland communities became widespread in Iowa under considerably hotter and drier conditions beginning sometime between 8,000 and 9,000 years ago. Deciduous trees, especially oaks, began to reappear in the state close to 3,000 years before present (Prior et al. 1982).

Pammel (1896) estimated that Iowa had 5 to 6 million acres of timberland shortly after settlement by Euro-American immigrants. In the 1930s, state forester G. B. MacDonald made a somewhat more refined estimate of the pre-Euro-American settlement forest area in the state by determining the proportion of each section line falling in "forest" or "scattered trees" from notes of the 1832–1859 Government Land Office survey. He reported a potential 6,680,926 acres of wooded land (in Crane and Olcott 1933; Thomson and Hertel 1981). As Thomson (1987) suggests, the distinction between different natural communities—in this case, between forest and savanna—is somewhat arbitrary due to natural biological overlap. The extent of what is presently considered forest was probably much less than the 6.7-million-acre figure, which most likely includes significant areas of savannas. A map based on MacDonald's study and a previous map published by Shimek (1899) show that potential forest land in Iowa (again, probably including what would presently be considered savanna) was concentrated in the eastern third of the state along the Iowa, Skunk, and Des Moines rivers and their major tributaries (Eilers 1982; Iowa State Planning Board 1935). This map is reproduced as figure 16.

As of 1974, only about 1.5 million acres of timberland in Iowa remained, more than a third of which was bottomland forest (USDA Forest Service 1980; Turner et al. 1981). A more recent survey has just been completed, and preliminary data indicate an increase of about .4 million acres in forest area in the state since the mid-1970s (Kingsley 1990).

The most dramatic reductions in forest area between the mid-1800s and the present occurred in eastern and southern Iowa (van der Linden and Farrar 1984). The earliest white settlers used the timber for lumber (for buildings and fences) and fuel. Before the turn of the century, selective logging of high-quality hardwoods such as walnut and oak (a practice known as high-grading) became profitable, which contributed to the deterioration of the forest community. Fencing livestock into woodland pastures became common, further degrading forests. The western extension of the railroad (roughly dating from the late 1850s) also required a large volume of timber (Thomson and Hertel 1981).

Finally, the development of large agricultural implements resulted in the destruction of much of the remaining forest land, which was

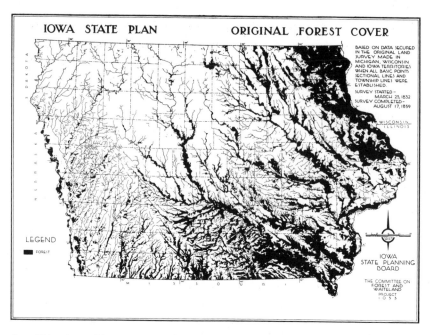

Figure 16. Iowa's original forest cover (Iowa State Planning Board 1935).

cleared for production of traditional agricultural crops from the late 1940s to the present (Thomson and Hertel 1981). Across the entire Midwest, forests which naturally appeared on choice level fertile sites were the first to be removed in favor of rowcrop production, while sites left as woodlands were those too steep, stony, wet, or otherwise unsuited for intensive agricultural use (Levenson 1981; Hightshoe 1984). This reduction of wooded areas on a regional scale has caused a decline in the biological diversity, productivity, and stability of the original forest communities of the area (Levenson 1981). In Iowa, the combined effects of selective logging, livestock damage to woodlands, and fragmentation of forests have resulted in remnant communities (second- and third-generation stands) which are species depauperate and genetically impoverished with respect to those species which remain (see Burgess and Sharpe 1981; Lovejoy and Oren 1981).

In addition to clearing for traditional agricultural uses, threats to remaining forest areas in Iowa include highway and utility corridor development, large reservoir impoundments, surface mining, and urban, suburban, and industrial expansion (Hightshoe 1984). Presently, however, there is a great deal of interest in tree planting in Iowa due to current environmental concerns and favorable cost-share programs through federal and state agencies. In many cases, establishment of trees is seen as a way to introduce permanent vegetative cover on a site, but as recently established plantations mature and form closed canopies, there may be opportunities to encourage development of mixed species stands and to introduce understory vegetation where necessary for restoration of more authentic forest communities.

TYPES OF FORESTS IN IOWA

Forests are areas with a relatively dense and extensive growth of woody plants which are at least 20 feet in height growing closely together (Society of American Foresters 1983; Spurr and Barnes 1980). Layers of vegetation beneath the canopy of dominant (overstory) trees are usually distinguishable (fig. 17), including understory trees, lianas (woody vines), shrubs, and herbaceous (nonwoody) ground-layer vegetation.

A number of deciduous trees reach the western limit of their natural range in eastern Iowa (van der Linden and Farrar 1984; Diekelmann and Schuster 1982). Aikman and Smelser (1938) referred to "the gradual dropping out of species as one proceeds westward across Iowa" (also true as one proceeds northward), which is especially noticeable for the overstory trees. Trees native to eastern Iowa include five species of hickory, ten species of oak, three species of ash, three species of birch, and five species of maple (van der Linden and Farrar 1984). Relatively few of these trees are naturally present west of the Des Moines River valley, as the climate becomes gradually colder, windier, and drier going from southeast to northwest across the state.

Iowa's major woodland types have been described according to forest community composition, and several are named for the predominant overstory species present. These communities recur predictably under certain soil moisture conditions and in specific physiographic positions. Most natural forest stands in the state can be placed in one of six communities: oak-hickory, oak-basswood, bot-

Figure 17. Layers of vegetation in an oak-hickory forest, from overstory trees to ground-layer herbs and grasses.

tomland hardwoods, riparian, northern conifer and hardwoods, and oak-cedar glade (Graham and Glenn-Lewin 1982; van der Linden and Farrar 1984). An abbreviated list of overstory, understory, shrub, and herbaceous ground-layer species usually present is given in the short description of each forest community that follows.

OAK-HICKORY

The oak-hickory forest community occurs throughout Iowa generally on dry upland sites and on south- and west-facing slopes (fig. 18). In the southeastern and eastern parts of the state, a large number of species may be present among overstory trees. Overstory species in extreme southeastern Iowa may include white oak (*Quercus alba*), black oak (*Quercus velutina*), post oak (*Quercus stellata*), shingle oak (*Quercus imbricaria*), chinkapin oak (*Quercus muehlenbergii*), bur oak (*Quercus macrocarpa*), blackjack oak (*Quercus marilandica*), northern pin oak (*Quercus ellipsoidalis*), shagbark hickory (*Carya ovata*), white ash (*Fraxinus americana*), black cherry (*Prunus serotina*), quaking aspen (*Populus tremuloides*), and big-

Figure 18. An upland oak-hickory forest on a south-facing slope in Story County's greenbelt along the Skunk River.

tooth aspen (*Populus grandidentata*) (Davidson 1960; van der Linden and Farrar 1984). In eastern Iowa, overstory species include white oak, black oak, chinkapin oak, red oak (*Quercus rubra*), bur oak, northern pin oak, shagbark hickory, white ash, black cherry, and the aspens. The dominant species in oak-hickory forests over much of the rest of Iowa are white oak, red oak, bur oak, and shagbark hickory in the central part of the state and bur oak in the western part of the state (Niemann and Landers 1974; Aikman and Smelser 1938). Understory trees in these stands may include ironwood (*Ostrya virginiana*), chokecherry (*Prunus virginiana*), hackberry (*Celtis occidentalis*), red mulberry (*Morus rubra*), serviceberry (*Amelanchier arborea*), American and slippery elms (*Ulmus americana* and *U. rubra*), and saplings of overstory trees (van der Linden and Farrar 1984; Davidson 1960). Shrubs are most abundant where the overstory canopy is relatively open and often include nannyberry (*Viburnum lentago*), prickly ash (*Zanthoxylum americanum*), prickly gooseberry (*Ribes cynosbati*), American hazelnut (*Corylus ameri-*

cana), smooth sumac (*Rhus glabra*), and gray and red osier dogwoods (*Cornus foemina* and *C. stolonifera*). Herbaceous ground-layer vegetation may include sedges (*Carex pensylvanica* and *C. muhlenbergii*), Solomon's seal (*Polygonatum canaliculatum*), goldenrods (*Solidago flexicaulis* and *S. ulmifolia*), bedstraw (*Galium concinnum*), wood anemone (*Anemone quinquefolia*), hog peanut (*Amphicarpaea bracteata*), and woody vines such as Virginia creeper (*Parthenocissus quinquefolia*) and grapes (*Vitus riparia* and *V. vulpina*) (Aikman and Smelser 1938; Niemann and Landers 1974; Davidson 1960).

The oak savanna (forest-grassland) is a distinct type of community with scattered large trees. At one time, oak savanna was probably a major natural community across the midwest, although few remnants of this ecosystem persisted long after Euro-American settlement (Packard 1988). Large areas of north-central, northeast, and south-central Iowa may have been covered by oak savannas, characterized by a relatively sparse overstory of bur and white oaks and an understory of unique grasses and forbs (Packard 1988; Roosa 1982). There is some evidence that the savanna, similar to the prairie, is a fire-dependent ecosystem and that many savanna understories were invaded by woody vegetation and gradually became forests following white settlement and suppression of fire (White 1983; Anderson and Brown 1986; Dorney and Dorney 1989). Former savannas that have become forests are often distinguishable by the presence of a small number of very large, branchy white and bur oak trees, with a much younger understory population of a number of species. Although some understory species have been identified in neighboring states with the help of historical documents (e.g., in Illinois, see Packard 1988), techniques for savanna reconstruction are not yet well defined.

OAK-BASSWOOD

The oak-basswood community occurs throughout Iowa on relatively moist but well-drained uplands, on protected north- and east-facing slopes, and on upper terrace levels in large stream valleys (fig. 19). In eastern Iowa, species dominant in the overstory include red oak, sugar maple (*Acer saccharum*), black maple (*Acer nigrum*), and basswood (also known as linden, *Tilia americana*) (van der Linden

Figure 19. Relatively dense shade in the understory of an oak-basswood forest community in Story County. Red oak, black maple, and basswood are the major overstory species in this stand.

and Farrar 1984; Davidson 1960). Proceeding from east to west across Iowa, sugar maples gradually disappear toward central Iowa and black maples disappear somewhat further west. Thus, the proportional importance of the red oak and basswood components in these stands increases going from east to west. Prior to the rapid spread of Dutch elm disease, American elm was also a significant overstory component of this forest community. Less numerous overstory trees may be white oak, shagbark hickory, bitternut hickory (*Carya cordiformis*), black walnut (*Juglans nigra*), butternut (*Juglans cinerea*), white ash, and black ash (*Fraxinus nigra*) (van der Linden and Farrar 1984; Davidson 1960). Understory trees are sparse but may include ironwood, black and sugar maple saplings, American hornbeam (*Carpinus caroliniana*), slippery elm, and black cherry. Among shrubs which may be present are wahoo (*Euonymous atropurpureus*), blackberry (*Rubus allegheniensis*), bladdernut (*Staphylea trifolia*), serviceberry, dogwoods, and witch hazel (*Hamamelis virginiana*). Ground-layer vegetation may include hepatica (*Hepatica acutifolia*), jack-in-the-pulpit (*Arisaema triphyllum*), false spike-nard (*Smilacina racemosa*), several species of sedges, wild sarsaparilla (*Aralia nudicaulis*), bloodroot (*Sanguinaria canadensis*), fragile fern (*Cystopteris fragilis*), Bishop's cap (*Mitella nuda*), trout lily (*Erythronium albidum*), dutchman's breeches (*Dicentra cucullaria*), and toothwort (*Dentaria* spp.) (van der Linden and Farrar 1984; Eilers n.d.).

BOTTOMLAND HARDWOODS

The bottomland hardwoods community occurs on floodplains and low-lying terraces in the larger stream valleys of Iowa (fig. 20; van der Linden and Farrar 1984). Overstory species present throughout the state include silver maple (*Acer saccharinum*), green ash (*Fraxinus pennsylvanica*), hackberry, black walnut, and cottonwood (*Populus deltoides*). In eastern Iowa, river birch (*Betula nigra*) is also an important overstory component. American elm was formerly dominant in this community as well. Other less common species which may be present in the overstory or as understory trees include sycamore (*Platanus occidentalis*), Kentucky coffee-tree (*Gymnocladus dioica*), honey locust (*Gleditsia triacanthos*), black willow (*Salix nigra*), peachleaf willow (*Salix amygdaloides*), bitternut hickory, shellbark hickory (*Carya laciniosa*), pin oak (*Quercus palustris*), rock elm (*Ulmus thomasii*), butternut, black ash, and, in southern and eastern Iowa, pecan (*Carya illinoensis*), shingle oak, swamp white oak (*Quercus bicolor*), and pawpaw (*Asimina triloba*) (van der Linden and Farrar 1984; Niemann and Landers 1974; Davidson 1960). Overstory species may be somewhat segregated according to slight changes in elevation and changes in soil texture: cottonwood, walnut, hackberry, sycamore, green ash, and silver maple appear on higher, better-drained coarser sediments, and willow and elms occur on low-lying finer-textured soils (Smith and Linnartz 1980). Except in areas that are frequently flooded, the understory is usually dense, with saplings of overstory trees, shrubs, and woody vines. Species such as wahoo, indigo bush (*Amorpha fruticosa*), button-bush (*Cephalanthus occidentalis*), chokecherry, elderberry (*Sambucus* spp.), dogwoods, grapes, Virginia creeper, and poison ivy (*Rhus radicans*) may be present. Herbaceous vegetation may include kidneyleaf buttercup (*Ranunculus arbortivus*), bristly butter-

Figure 20. A bottomland forest along the Skunk River in Story County. Common species present include black walnut, silver maple, ironwood, and green ash.

cup (*Ranunculus pensylvanicus*), impatiens (*Impatiens capensis* and *I. pallida*), nettle (*Urtica dioica*), wood nettle (*Laportea canadensis*), goldenrods (*Solidago* spp.), Virginia waterleaf (*Hydrophyllum virginianum*), and vervain (*Verbena* spp.) (Niemann and Landers 1974; Davidson 1960).

RIPARIAN FORESTS

The riparian forest community occurs as a narrow belt along streams, mudflats, sandbars, and lakeshores (fig. 21). Trees present in the riparian community include cottonwood, silver maple, box elder (*Acer negundo*), river birch, sandbar willow (*Salix exigua*), black willow, and peachleaf willow, as well as some of the bottomland hardwood species. Again, dominant species are usually segregated by microtopographical differences as discussed with respect to bottomland hardwoods. Understory trees and shrubs may include plum (*Prunus americana*), chokecherry, elms, other species of willows (silky willow, *Salix sericea*; rigid willow, *S. lutea*; and prairie willow, *S. humilis*), wahoo, and elderberry (Graham and Glenn-

Figure 21. Riparian forest along the Des Moines River, Dolliver Memorial State Park, Webster County.

Lewin 1982; van der Linden and Farrar 1984). Ground-layer vegetation may include species which are found in bottomland forests as well as those typical of wetlands such as smartweed (*Polygonum* spp.), sedges (*Carex* spp.), and beggar-ticks (*Bidens comosa*).

NORTHERN CONIFERS AND HARDWOODS

The northern conifer and hardwood forest community occurs in a very specific microclimate on moist, usually steep north-facing slopes in northeastern Iowa (fig. 22). Overstory species generally include those of the oak-basswood forest community together with white pine (*Pinus strobus*), balsam fir (*Abies balsamea*), paper birch (*Betula papyrifera*), yellow birch (*Betula allegheniensis*), and mountain maple (*Acer spicatum*). Other overstory and understory trees which may be present are quaking aspen, bigtooth aspen, and black ash. Shrubs common in this forest community include speckled alder (*Alnus rugosa*), Canada yew (*Taxus canadensis*), highbush cranberry (*Viburnum opulus*), red elderberry (*Sambucus pubens*), and red osier dogwood. Herb and ground-layer vegetation includes some rare species and others that are at least rare in the state: red bane-

Figure 22. The northern conifer and hardwood forest in White Pine Hollow, Dubuque County. Cold air drainage into this area protects a relict population of white pine, sugar maple, aspen, and birch trees.

berry (*Actaea rubra*), spikenard (*Aralia racemosa*), northern bush honeysuckle (*Diervilla lonicera*), leatherwood (*Dirca palustris*), wild lily-of-the-valley (*Maianthemum canadense*), ginseng (*Panax quinquefolium*), and slender rock brake fern (*Cryptogramma stelleri*) (Davidson 1960; Eilers n.d.).

OAK-CEDAR GLADE

The oak-cedar limestone glade occurs on alkaline soils, thin soils over limestone bedrock, and sites where very little soil is present over bedrock, primarily in extreme eastern Iowa (fig. 23). The predominant tree species present are chinkapin oak and eastern red cedar (*Juniperus virginiana*) (Graham and Glenn-Lewin 1982). Shrub and ground-layer vegetation may include downy serviceberry, columbine (*Aquilegia canadensis*), hops (*Humulus lupulus*), moonseed (*Menispermum canadense*), black snake-root (*Sanicula canadensis*), bladder fern (*Cystopteris bulbifera*), starry campion (*Silene stellata*), sassafras (*Sassafras albidum*), goldenrods, pye-weeds (*Eupatorium* spp.), flowering spurge (*Euphorbia corollata*), false sunflower

Figure 23. An oak-cedar glade forest community on a steep slope in Allamakee County.

(*Heliopsis helianthoides*), crownbeard (*Verbesina alternifolia*), and ironweed (*Vernonia baldwinii*) (Davidson 1960; Eilers n.d.).

BENEFITS OF FOREST RESTORATION

Restoration of forest communities can provide multiple benefits, including protection and improvement of environmental quality, wildlife habitat, agricultural diversification, and sustained yields of marketable forest products. Additional benefits of woodland restoration accrue from social values such as preservation of the genetic resource; provision of natural areas for research, education, and recreation; and aesthetic values, especially those associated with landscape variety in a dominantly agricultural environment.

ENVIRONMENTAL QUALITY

Erosion control and water-quality protection are frequently cited among the important benefits of afforestation/reforestation efforts (e.g., Peterjohn and Correll 1984; Howell 1986; Wray 1987; van der Linden and Farrar 1984). Forest vegetation limits soil erosion by reducing rainfall impact via interception in the canopy, understory, and litter layers and by stabilizing the soil with a network of large, permanent roots (Howell 1986; Cubbage and Gunter 1987). Reduction of nonpoint source pollution by riparian forests (e.g., removal of sediments and excess fertilizer produced in agricultural watersheds) is a significant factor in protecting streams and rivers from excessive nutrient loading (Peterjohn and Correll 1984). Riparian forest vegetation also helps moderate water temperatures, provides habitat for insects (food chain support), and directly supplies detrital material to improve fish habitat (Oliver and Hinckley 1987).

WILDLIFE HABITAT

The role of forests in providing habitat for wildlife is especially critical in Iowa, where permanent cover for wildlife is otherwise very limited (Brown et al. 1978; Capel 1988). The structural complexity and diversity of food sources and cover common in temperate forests can support a variety of animal and bird species throughout the year (Johnson and Beck 1988; Brinson et al. 1981). Although all trees have some value to wildlife as potential nest or den sites (fig. 24), species such as oak, pine, cherry, dogwood, maple, sumac, hackberry, red cedar, and aspen provide particularly valuable foods, including seeds, fruits, foliage, buds, flowers, and twigs (van der Linden and Farrar 1984). Streamside (riparian) forests can provide important habitat and travel corridors for migration of both plant and animal species and are even more useful if they occur as links between larger forests which extend to adjacent uplands (MacClintock et al. 1977; Stauffer and Best 1980; Forman 1983).

Species of birds common in Iowa which primarily utilize forest or forest-edge habitat include the red-tailed hawk, wild turkey, American woodcock, great horned owl, barred owl, eastern bluebird (savanna-type woodlands), rose-breasted grosbeak, black-capped chickadee, eastern wood-pewee, tufted titmouse, white-breasted nuthatch, blue jay, northern oriole, ruby-throated hummingbird, and several species of woodpeckers, wrens, vireos, warblers, kinglets, and flycatchers. Bird species associated with riparian woodlands include many of those already mentioned as well as the green-backed heron, yellow-crowned night heron, wood duck, hooded merganser, turkey vulture, whippoorwill, and tree swallow (Dinsmore et al. 1984). A growing number of forest-dwelling bird species are encoun-

Figure 24. A great
horned owl tends her
nest in the cavity of a
cottonwood tree. Photo
by Linda Schultz.

tered much less frequently in the state than they were in the past due to direct loss of habitat (especially forest interior habitat) and fragmentation of the forests that remain. Among the species which may be most affected by decreases in habitat area are the oven-bird, red-shouldered hawk, Bell's vireo, sharp-shinned hawk, broad-winged hawk, long-eared owl, winter wren, black-and-white war-bler, and worm-eating warbler. Additional species which appear to require undisturbed riparian forest areas include the scarlet tanager, American redstart, wood thrush, and rufous-sided towhee (Dins-more 1981; Robbins 1979; Stauffer and Best 1980).

Many small mammals are still common in Iowa woodlands: the white-footed mouse, deer mouse, short-tailed shrew, least shrew, gray squirrel, fox squirrel, red squirrel, southern flying squirrel, er-mine, opossum, least weasel, eastern chipmunk, eastern cottontail, and red bat (Bowles 1981; Gottfried 1979). Larger mammals which use forest or forest-edge habitat include the raccoon, woodchuck, gray fox, and white-tailed deer. Forest-dwelling mammals which are threatened or have been extirpated from the state due to hunting and/or habitat destruction are Keen's myotis, Indiana bat, evening bat, red-backed vole, woodland vole, porcupine, gray wolf, black bear, wolverine, and lynx (Bowles 1981). Restoration of adequate forest areas in the state could encourage population growth for endangered and threatened species of wildlife that were formerly more common in Iowa.

SUSTAINED YIELD AND AGRICULTURAL DIVERSIFICATION

Managing forest areas for multiple uses and sustained yields of a variety of products has been the objective of professional foresters for decades. Even in woodlands designated primarily as natural areas, sustainable economic benefits from production of maple sap, fuelwood, and in some cases raw biomass or timber can be realized with limited disturbance to the forest area (Beattie et al. 1983). Growing trees in addition to traditional agricultural crops could promote much needed agricultural diversification and landscape variety and help curb excess production of grain commodities (Hightshoe 1984; Cubbage and Gunter 1987). Some very high quality timber species are native to Iowa, including black walnut and red oak (van der Linden and Farrar 1984). Several government agencies offer financial incentives for tree planting, particularly on lands susceptible to damage from intensive cultivation (e.g., Iowa Department of Natural Resources Resource Enhancement and Protection and the federal Conservation Reserve Program).

SOCIAL VALUES

Additional benefits of forest restoration include preservation of natural heritage values such as genetic and species diversity. Restored forest communities may provide opportunities for research and educational activities while preventing damage to naturally occurring but rare habitats. Restoration of forest communities can also enhance the aesthetic value of recreational areas. For areas not managed for specific forest products, maintenance costs are minimal once the overstory species are established.

FOREST ESTABLISHMENT

Structural complexity and successional dynamics—certain species tend to follow others over fairly long periods of time, determined to a certain extent by the relative ability of the species to reproduce under the canopy—in forest communities are such that complete forest restoration should be at least in part a natural process allowed to progress over a long period of time (Ashby 1987; Hightshoe 1984). The principal objective of forest restoration, according to Hightshoe (1984), should be to provide suitable habitat for reestablishment of forest community species (overstory, understory, and herbaceous plants) where they formerly existed and to manage restoration areas in ways that encourage maximum natural system biodiversity.

Howell (1986) outlines two general approaches which have been used in forest restoration: (1) duplication of a model forest community, with attention given to re-creating the relative abundance, age-class structure (overstory and understory), spacing, and distribution of the same species found in a natural forest occurring on a similar site, or (2) simulation of ecosystem functions such as nutrient retention, erosion control, structural diversity, or biomass production, which has usually involved a similar combination of life forms (overstory and understory layers) although exact species composition may not mimic natural forests in the area. The first approach is closer to the ideal for restoration, but even this can only approximate the original forest when applied in much of the Midwest because of the biological impoverishment of the model forest communities that remain. An underlying assumption of either approach is that assisting in the reproduction of forest structure will eventually result in the dynamics and functions provided by natural stands (Howell 1986).

To plan a forest restoration project, consideration should be given to site selection, species to be planted, site preparation, planting methods, sources of propagules, and maintenance of the forest.

SITE SELECTION

Iowa's forests at the time of Euro-American settlement were concentrated in the eastern, southeastern, and south-central parts of the state, in some small areas in extreme western Iowa, and along rivers and streams throughout much of the rest of the state (see fig. 16). Savannas may also have been widespread in north-central and eastern Iowa. Forest restoration per se is appropriate only in previously forested areas. Riparian habitats represent particularly valuable areas for forest restoration because of their importance as buffer strips and wildlife corridors.

Soils which formed under the influence of forest vegetation can serve as indicators for potential forest restoration sites in areas where forest vegetation no longer persists. Many midwestern soils which are classified as Alfisols (USDA 1975) for the most part formed under mixed hardwood forests (Rust 1983). The distinctive characteristics of these soils include relatively shallow, light-colored surface horizons (especially compared to prairie-influenced soils), a bleached zone from which material (clay, organic matter) has been moved to a lower layer, and a subjacent layer where this material is deposited and accumulates. County soil surveys contain soil maps, descriptions of soil map units, and other information that may be helpful in identifying potential forest restoration sites where trees are no longer present. Local Soil Conservation Service personnel may be able to assist in evaluating site suitability. Forest restoration efforts are also appropriate in degraded forest areas where some trees remain but understory species are missing or sparse (Hightshoe 1984).

The type of vegetation present on a potential restoration site will affect the amount of time and effort required to reestablish a complete forest community. In areas with a natural forest overstory already in place, it is usually possible to manipulate the existing canopy, adding or removing individuals or species and improving the understory habitat for seedlings, shrubs, and herbaceous vegetation (Howell 1986; Hightshoe 1984). Generally, restoration is faster and easier with an established canopy or at least scattered large trees in place to provide suitable microclimatic conditions for the introduction of understory species. More patience and intensive effort are required to establish forest communities where no overstory trees are present. Depending on the species appropriate to the site, restoration may take as long as forty to one hundred years.

Other factors to consider in selecting a site are the size and shape of the potential restoration site, proximity to other woodlands or areas of permanent vegetation, potential for weed competition during seedling establishment, location of natural firebreaks, accessibility, and the necessity to exclude domestic livestock from the area.

Although edge habitats are typically very species diverse and attract a variety of wildlife, to attract and protect some forest interior species it may be necessary to restore areas of at least 10 to 20 acres (Levenson 1981; Anderson 1990), but the ideal area is much larger.

Irregularly shaped boundaries (i.e., not rectangular plantations) impart a more natural appearance as the trees mature. Connections with other areas of permanent vegetation, especially other forest areas, act as corridors that are conducive to migrations of both plant and animal species and can make the restored forest particularly attractive to wildlife (Opdam et al. 1985; MacClintock et al. 1977; Merriam 1989). To facilitate fire protection, it may be advantageous to select a site with firebreaks "built in," such as roads, streams, ditches (at least 18 feet wide), or cultivated fields (Wray 1988). If management for amenities other than wildlife is likely (e.g., sugarbush, fuelwood, or timber), access roads will probably be needed. In addition, accessibility of the area to farm machinery may be important for large restoration efforts in order to facilitate machine planting. Exclusion of livestock from the forest restoration area (although perhaps not for savanna restoration areas) is absolutely necessary, so placement of the site with respect to areas used by domestic animals should also be considered. Proper fencing to prevent damage by nearby livestock is essential.

SPECIES TO BE PLANTED

Native species from appropriate forest community types should be carefully matched to site characteristics, especially soil moisture conditions, topography (slope and aspect), and location in the state (Diekelmann and Schuster 1982; Howell 1986; Wray and Farris 1989). As indicated earlier for the natural forest types that occur in Iowa, many more species can be considered for planting in eastern Iowa than in the western half of the state. Although both conifers and deciduous trees are suitable for planting, only five conifer species were native to the state at the time of Euro-American settlement, and four of these occur naturally only in the northeast corner of Iowa (van der Linden and Farrar 1984). A partial list of forest community species and their typical growth form, growth rate, distribution in the state, soil requirements, and ability to survive and reproduce in the shaded understory is given in table 2. Consultation with a professional (Department of Natural Resources district forester, extension forester, consulting forester, county conservation board biologist, or Soil Conservation Service personnel) is recommended to assist in species selection. In addition, as discussed with respect to prairie reconstruction, every attempt should be made to use local genotypes for forest restoration.

Table 2. Selected Species Suitable for Forest Restoration in Iowa

Species	Form	Growth Rate	Distribution	Soil Requirements	Tolerance
Community: Oak-Hickory					
Red oak	large tree	mod. fast	throughout	moist, well-drained	intermediate
White oak	large tree	slow	all except ext. w	dry upland	intermediate
Black oak	large tree	moderate	s and e half	well-drained	intermediate
Post oak	small tree	slow	extreme s	dry upland	intolerant
Shingle oak	small tree	slow	extreme se	moist to dry	intolerant
Chinkapin oak	small tree	slow	extreme se	well-drained, calcareous upland	intolerant
Bur oak	large tree	slow	throughout	dry upland	intolerant
Blackjack oak	small tree	slow	extreme s	dry upland	intolerant
N. pin oak	smalll tree	mod. fast	n half	moist to dry	v. intolerant
Pin oak	small tree	mod. fast	s and e	moist to dry	v. intolerant
Shagbark hickory	large tree	v. slow	all except ext. nw	dry upland	intermediate
White ash	large tree	mod. fast	all except ext. nw	moist to dry	tolerant
Black cherry	large tree	mod. fast	throughout	moist upland	intolerant
Quaking aspen	large tree	v. fast	e	dry upland	intolerant
Bigtooth aspen	large tree	v. fast	e	dry upland	intolerant
Ironwood	small tree	v. slow	throughout	ubiquitous	v. tolerant
Chokecherry	small tree	fast	throughout	moist to dry	intermediate
Hackberry	small tree	mod. fast	throughout	moist, well-drained	intermediate
Red mulberry	small tree	moderate	s and e	moist, sheltered	intermediate
Serviceberry	small tree	moderate	e, s, and central	moist, well-drained	v. tolerant
American elm	small tree	fast	throughout	ubiquitous	intermediate
Slippery elm	small tree	fast	throughout	ubiquitous	intermediate
Nannyberry	shrub	fast	ne and n	moist upland	intermediate
Prickly ash	shrub	fast	throughout	dry upland	intolerant
Prickly gooseberry	shrub	fast	throughout	dry upland	?
American hazelnut	shrub	fast	throughout	dry upland	?
Smooth sumac	shrub	fast	throughout	dry upland	?
Gray dogwood	shrub	fast	n	moist, sheltered	tolerant
Virginia creeper	liana	(probably requires presence of overstory to create suitable habitat)			
Grapes	liana	(probably requires presence of overstory to create suitable habitat)			
Sedges	groundlayer	(probably requires presence of overstory to create suitable habitat)			

Table 2. *(continued)*

Species	Form	Growth Rate	Distribution	Soil Requirements	Tolerance
Solomon's seal	groundlayer	(probably requires presence of overstory to create suitable habitat)			
Goldenrod	groundlayer	(probably requires presence of overstory to create suitable habitat)			
Bedstraw	groundlayer	(probably requires presence of overstory to create suitable habitat)			
Wood anemone	groundlayer	(probably requires presence of overstory to create suitable habitat)			
Hog peanut	groundlayer	(probably requires presence of overstory to create suitable habitat)			

Community: Oak-Basswood

Species	Form	Growth Rate	Distribution	Soil Requirements	Tolerance
Red oak	large tree	mod. fast	throughout	moist, well-drained	intermediate tolerance
Sugar maple	large tree	moderate	e	moist, well-drained	v. tolerant
Black maple	large tree	mod. slow	e and central	moist, well-drained	v. tolerant
Basswood	large tree	moderate	throughout	moist, well-drained	tolerant
White oak	large tree	slow	all except ext. w	dry upland	intermediate
Shagbark hickory	large tree	slow	all except ext. nw	dry upland	intermediate
Bitternut hickory	large tree	slow	all except ext. nw	moist	intermediate
Black walnut	large tree	fast	throughout	sheltered, moist, well-drained	intolerant
Butternut	large tree	fast	e half	sheltered, moist, well-drained	intolerant
White ash	large tree	mod. fast	all except ext. nw	moist to dry	tolerant
Black ash	large tree	fast	e and ne two-thirds	wet	intolerant
Ironwood	small tree	v. slow	throughout	none	v. tolerant
American hornbeam	small tree	slow	ext. e	moist	v. tolerant
Slippery elm	small tree	moderate	throughout	moist	intermediate
Black cherry	small tree	mod. fast	throughout	moist	intolerant
Wahoo	shrub	moderate	throughout	moist	tolerant
Blackberry	shrub	moderate	throughout	moist	tolerant
Bladdernut	shrub	moderate	throughout	moist	tolerant
Serviceberry	shrub	moderate	e, s, and central	moist	v. tolerant
Dogwoods	shrub	fast	all except ext. nw	moist to dry	tolerant
Witch hazel	shrub	moderate	extreme ne	moist	v. tolerant
Hepatica	groundlayer	(probably requires presence of overstory to create suitable habitat)			

Table 2. *(continued)*

Species	Form	Growth Rate	Distribution	Soil Requirements	Tolerance
Jack-in-the-pulpit	groundlayer	(probably requires presence of overstory to create suitable habitat)			
False spikenard	groundlayer	(probably requires presence of overstory to create suitable habitat)			
Sedges	groundlayer	(probably requires presence of overstory to create suitable habitat)			
Sarsaparilla	groundlayer	(probably requires presence of overstory to create suitable habitat)			
Bloodroot	groundlayer	(probably requires presence of overstory to create suitable habitat)			
Fragile fern	groundlayer	(probably requires presence of overstory to create suitable habitat)			
Bishop's cap	groundlayer	(probably requires presence of overstory to create suitable habitat)			
Trout lily	groundlayer	(probably requires presence of overstory to create suitable habitat)			
Dutchman's breeches	groundlayer	(probably requires presence of overstory to create suitable habitat)			
Toothwort	groundlayer	(probably requires presence of overstory to create suitable habitat)			

Community: Bottomland Hardwoods

Species	Form	Growth Rate	Distribution	Soil Requirements	Tolerance
Silver maple	large tree	v. fast	throughout	wet to moist	intermediate tolerance
Green ash	large tree	fast	throughout	moist	intolerant
Hackberry	large tree	mod. fast	throughout	moist	intermediate
Black walnut	large tree	fast	throughout	moist	intolerant
Cottonwood	large tree	v. fast	throughout	moist	intolerant
River birch	large tree	fast	ext. e and ne	moist	intolerant
Sycamore	large tree	v. fast	s half	moist	intermediate
Kentucky coffee-tree	large tree	slow	s and central	moist	intolerant
Honey locust	large or small tree	fast	all except ext. n	moist	intolerant
Black willow	large or small tree	v. fast	all except ext. nw	wet	v. intolerant
Peachleaf willow	large or small tree	fast	throughout	wet	v. intolerant
Bitternut hickory	large or small tree	slow	all except ext. nw	moist	intermediate

Table 2. *(continued)*

Species	Form	Growth Rate	Distribution	Soil Requirements	Tolerance
Shellbark hickory	large or small tree	slow	s	moist	intermediate
Pin oak	large or small tree	mod. fast	s and e	wet to moist	v. intolerant
Rock elm	large or small tree	moderate	throughout	moist	intermediate
Butternut	large or small tree	fast	e half	moist, well-drained	intolerant
Black ash	large or small tree	fast	e and ne two-thirds	wet	intolerant
Pecan	large or small tree	slow	ext. e and se	moist	intolerant
Shingle oak	large or small tree	slow	ext. se	moist	intolerant
Swamp white oak	large or small tree	moderate	e half	wet	intermediate
Pawpaw	large shrub	slow	ext. se	moist	tolerant
Wahoo	shrub	moderate	s and e	moist to wet	tolerant
Indigo bush	shrub	moderate	throughout	moist	tolerant
Buttonbush	shrub	moderate	throughout	moist	tolerant
Chokecherry	shrub	fast	throughout	dry to moist	intermediate
Elderberry	shrub	fast	throughout	moist	intermediate
Dogwoods	shrub	mod. fast	all except ext. nw	moist to dry	tolerant
Grapes	liana	(probably requires presence of overstory to create suitable habitat)			
Virginia creeper	liana	(probably requires presence of overstory to create suitable habitat)			
Poison ivy	liana	(probably requires presence of overstory to create suitable habitat)			
Buttercups	groundlayer	(probably requires presence of overstory to create suitable habitat)			

Table 2. *(continued)*

Species	Form	Growth Rate	Distribution	Soil Requirements	Tolerance
Impatiens	groundlayer	(probably requires presence of overstory to create suitable habitat)			
Nettles	groundlayer	(probably requires presence of overstory to create suitable habitat)			
Goldenrods	groundlayer	(probably requires presence of overstory to create suitable habitat)			
Vervain	groundlayer	(probably requires presence of overstory to create suitable habitat)			
Virginia waterleaf	groundlayer	(probably requires presence of overstory to create suitable habitat)			

Community: Riparian Forest

Species	Form	Growth Rate	Distribution	Soil Requirements	Tolerance
Cottonwood	large tree	v. fast	throughout	wet to moist	intermediate
Silver maple	large tree	v. fast	throughout	moist	intolerant
Box elder	large tree	v. fast	throughout	wet	v. intolerant
River birch	large tree	fast	ext. e and ne	moist	intolerant
Sandbar willow	large tree	fast	throughout	wet	v. intolerant
Black willow	large tree	v. fast	all except ext. nw	wet	v. intolerant
Peachleaf willow	small tree	fast	throughout	wet	v. intolerant
Plum	small tree	fast	throughout	moist, well-drained	intolerant
Chokecherry	small tree	fast	throughout	wet to dry	intermediate
Elms	small tree	fast	throughout	wet to dry	intermediate
Silky willow	large shrub	fast	throughout	wet	intolerant
Rigid willow	large shrub	fast	throughout	wet	intolerant
Prairie willow	large shrub	fast	throughout	wet	intolerant
Wahoo	shrub	moderate	throughout	moist	tolerant
Elderberry	shrub	fast	throughout	moist	intermediate
Smartweed	groundlayer	(probably requires presence of overstory to create suitable habitat)			
Sedges	groundlayer	(probably requires presence of overstory to create suitable habitat)			
Beggar-ticks	groundlayer	(probably requires presence of overstory to create suitable habitat)			

Community: Northern Conifer and Hardwood

Species	Form	Growth Rate	Distribution	Soil Requirements	Tolerance
Red oak	large tree	mod. fast	extreme ne	moist, sheltered, well-drained	intermediate
Sugar maple	large tree	moderate	extreme ne	moist, sheltered, well-drained	v. tolerant
Black maple	large tree	mod. slow	extreme ne	moist, sheltered, well-drained	v. tolerant

Table 2. *(continued)*

Species	Form	Growth Rate	Distribution	Soil Requirements	Tolerance
Basswood	large tree	moderate	extreme ne	moist, sheltered, well-drained	tolerant
White oak	large tree	slow	extreme ne	moist, sheltered, well-drained	intermediate
White pine	large tree	moderate	extreme ne	moist, sheltered, well-drained	intermediate
Balsam fir	large tree	slow	extreme ne	moist, sheltered, well-drained	v. tolerant
Paper birch	large tree	moderate	extreme ne	moist, sheltered, well-drained	v. intolerant
Yellow birch	large tree	moderate	extreme ne	moist, sheltered, well-drained	tolerant
Mountain maple	small tree	slow	extreme ne	moist, sheltered, well-drained	v. tolerant
Quaking aspen	small tree	v. fast	extreme ne	moist, sheltered, well-drained	intolerant
Bigtooth aspen	small tree	v. fast	extreme ne	moist, sheltered, well-drained	intolerant
Black ash	small tree	fast	extreme ne	moist, sheltered, well-drained	intolerant
Speckled alder	shrub	fast	extreme ne	moist, sheltered, well-drained	v. intolerant
Canada yew	shrub	moderate	extreme ne	moist, sheltered, well-drained	intermediate
Highbush cranberry	shrub	fast	extreme ne	moist, sheltered, well-drained	intermediate
Red elderberry	shrub	fast	extreme ne	moist, sheltered, well-drained	intermediate
Red osier dogwood	shrub	fast	extreme ne	moist, sheltered, well-drained	tolerant
Red baneberry	groundlayer	(probably requires presence of overstory to create suitable habitat)			
Spikenard	groundlayer	(probably requires presence of overstory to create suitable habitat)			
Northern bush honeysuckle	groundlayer	(probably requires presence of overstory to create suitable habitat)			

Table 2. *(continued)*

Species	Form	Growth Rate	Distribution	Soil Requirements	Tolerance
Leatherwood	groundlayer	(probably requires presence of overstory to create suitable habitat)			
Lilly-of-the-valley	groundlayer	(probably requires presence of overstory to create suitable habitat)			
Ginseng	groundlayer	(probably requires presence of overstory to create suitable habitat)			
Rock brake fern	groundlayer	(probably requires presence of overstory to create suitable habitat)			

Community: Oak-Cedar Glade

Species	Form	Growth Rate	Distribution	Soil Requirements	Tolerance
E. red cedar	tree	slow	throughout	dry upland	intermediate for juvenile, intolerant when mature
Chinkapin oak	tree	slow	ext. e	dry upland	intolerant
Serviceberry	large shrub	moderate	throughout	moist to dry	v. tolerant
Three-awn grasses	groundlayer	(may require presence of overstory to create suitable habitat)			
Bladder fern	groundlayer	(may require presence of overstory to create suitable habitat)			
Aureolaria	groundlayer	(may require presence of overstory to create suitable habitat)			
Field milkwort	groundlayer	(may require presence of overstory to create suitable habitat)			
Columbine	groundlayer	(may require presence of overstory to create suitable habitat)			
Hops	groundlayer	(may require presence of overstory to create suitable habitat)			
Moonseed	groundlayer	(may require presence of overstory to create suitable habitat)			
Black snake-root	groundlayer	(may require presence of overstory to create suitable habitat)			
Starry campion	groundlayer	(may require presence of overstory to create suitable habitat)			
Sassafras	groundlayer	(may require presence of overstory to create suitable habitat)			
Goldenrods	groundlayer	(may require presence of overstory to create suitable habitat)			
Pye-weeds	groundlayer	(may require presence of overstory to create suitable habitat)			
Flowering spurge	groundlayer	(may require presence of overstory to create suitable habitat)			
False sunflower	groundlayer	(may require presence of overstory to create suitable habitat)			
Crownbeard	groundlayer	(may require presence of overstory to create suitable habitat)			
Ironweed	groundlayer	(may require presence of overstory to create suitable habitat)			

Note: Tolerance refers to ability of the species to survive and reproduce in the shaded understory.

Sources: van der Linden and Farrar 1984; Diekelmann and Schuster 1982; Hightshoe 1978; Davidson 1960.

SITE PREPARATION

The amount of effort required for site preparation will depend on soil and vegetation characteristics of the site prior to restoration. For sites with an existing forest canopy or sites with light-textured soil without dense weed or sod cover, very little site preparation may be necessary (Howell 1986; Wray 1988). On many sites in Iowa, however, sod or weed competition will be a serious problem in establishing tree seedlings, and starting weed control before planting trees is strongly recommended (Merritt 1980). Existing biennial or perennial sod or weeds can be removed by cultivating (disking, hoeing, rototilling, or harrowing) on the contour or by using a herbicide such as glyphosate (Roundup) in the fall before spring planting. Herbicides should be applied by certified individuals strictly in accordance with label specifications. All vegetation in strips or circles 2 to 4 feet wide should be removed where seedlings are to be planted (Wray 1988). Herbicide use is specifically recommended for fall kill of vegetation on slopes, leaving standing dead material in place to protect against erosion. A prescribed burn could also be used to remove existing vegetation under certain circumstances. Use of fire as a site preparation technique requires strict adherence to professional guidelines and will probably require an open burning permit (consult local fire departments, Department of Natural Resources biologists, or county conservation board biologists).

PLANTING METHODS

The first priority in forest restoration is to establish a canopy of overstory trees if one is not already present on the site. Howell (1986) outlines several methods which have been used for woodland restoration, and the most practical methods are listed here.

1. Planting canopy trees at ultimately desired densities and proportions; mulching or spraying to control weeds in the immediate vicinity of the trees; encouraging natural invasion by mid- and understory species as shade develops.

2. Planting trees at less than desired densities; as canopy develops, planting more trees and understory species (gives a natural appearance).

3. Planting trees at greater than desired densities to control unwanted vegetation on site, then thinning or allowing to self-thin;

adding understory species later (in most cases it is necessary to intervene and do some thinning).

4. Planting short-lived or fast-growing trees or tall shrubs to capture the site; underplanting slow-growing shade-tolerant trees that will eventually become dominants (more or less accelerated natural succession).

5. Not planting trees and instead allowing woody species to naturally invade (assuming a nearby seed source); selectively removing undesirable species in under- and overstory (this method may take several human lifetimes to establish a forest).

The most successful approach in any case will depend on the ecology of the site and the target community, the speed with which results are needed, and cost factors (Howell 1986).

Seeds, cuttings, seedlings, or larger trees can be used as planting stock for forest canopy establishment. Planting methods for each are summarized in the following paragraphs.

PLANTING SEED. It is possible to collect and condition seed for tree species native to Iowa, and information on collecting seed and seed conditioning for some species is given in table 3. However, direct seeding (planting seed directly on the restoration site), particularly for hardwoods (especially in the fall), is generally not recommended in the Midwest due to heavy rodent damage and intense weed competition (Merritt 1980; Forbes 1971; Beattie et al. 1983). Moderate success may be achieved with direct seeding only on relatively large areas (larger than 5 acres) and on very open sites (Wray 1990).

Seed should only be collected from healthy trees that bear large quantities of seed (Wray 1986). Many species require artificial stratification (cold storage) if seed is collected in the fall and is to be planted in the spring. Storing seed between moist layers of sand or peat moss at 40 degrees for two to four months is generally adequate stratification. More complete information on seeds and seed conditioning is given in *Seeds of Woody Plants in the United States* (USDA 1974).

Seed can be planted on a relatively large scale using machinery (e.g., a corn planter) adapted for the size of seeds to be planted. For seeds that are especially attractive to wildlife, such as acorns or walnuts, planting as many as 2,000 seeds per acre may produce a stand of about one hundred trees per acre.

Wray (1986) recommends germinating seeds and culturing them

Table 3. Seed Collection and Seed Conditioning for Some Forest Tree Species of Iowa

Species	Collect Ripe Seed	Stratify (Days)	Plant (Non-stratified)	Depth Seed (In.)	Density (Seeds/ Sq. Ft.)
Ashes	Aug.–Sept.	60	fall	0.5	10–15
Fir	Aug.–Sept.	30	fall	0.5	50
Basswood	Sept.–Oct.	90	fall	0.25–0.5	30
Birch	Aug.–Sept.	30–80	fall	0.25	30
Butternut	Sept.–Oct.	90–120	fall	1–2	10–15
Red cedar	Sept.–Nov.	30–120	fall	0.25	30–50
Cherries	July–Sept.	120	fall	0.5	30
Cottonwood	May–June		spring	0	20
Box elder	Sept.–Oct.	60–90	fall	0.5	20
Hackberry	Sept.–Oct.	60–90	fall	0.5	20
Hickories	Sept.–Oct.	90	fall	1	10
Red, silver maple	Mar.–May		spring	0.5	20
Sugar, black maple	Sept.–Oct.	60	fall	0.5	10–15
White oak	Sept.–Oct.		fall	0.75	10
Red, black oak	Sept.–Oct.	30–90	fall	0.75	10
Pine	Aug.–Sept.	0–60[a]	spring/fall	0.25	50
Walnut	Sept.–Oct.	90–150	fall	1–2	10

Source: Wray 1986.

[a] Cones should be collected after seeds are mature but before cones open. Seeds lie at the base of the cone scales. Cones can be opened using mild heat (less than 120° for 4 hours). Unstratified pine seeds can be planted in the fall with reasonable success.

for one or two years in a raised seedbed, then transplanting the one- or two-year-old seedlings onto the planting site. The seedbed should be located on a well-drained site that receives plenty of sunlight, is near an adequate source of water, and is in an area that can be protected from animals and livestock. Seedlings can be shaded (snow fence, wooden laths, or commercially available shadecloth can be used) when very young. Weeds should be removed from the seedbed frequently. Seedlings should be watered twice a week for the first two months and once a week thereafter if rainfall is inadequate. Seedlings can be transplanted using the directions given for planting bare-root seedlings in the section that follows.

PLANTING TREE SEEDLINGS. Many tree species, especially hard-woods, are most successfully established in the Midwest when planted as one- or two-year-old seedlings (Smith and Linnartz 1980; Merritt 1980). Larger container or liner-grown stock may be used on difficult sites, but transplanting large stock is very labor-intensive and is probably not feasible for any large-scale planting effort. For plantings much larger than an acre, it is probably most practical to use one- or two-year-old bare-root seedlings which are available from many commercial nurseries. Although planting should be done as soon as possible after seedlings arrive, if planting stock is shipped in quantity for a large project it may be necessary to cold store or "heel in" the seedlings until the time of planting (this is recommended if seedlings must be held a week or more; see Forbes 1971). Planting may be done as soon as danger of frost is past, roughly March 20 (perhaps later for conifers, depending on soil temperatures) in southern Iowa and between April 1 and May 15 throughout the rest of the state. Fall planting has been done with some success and may be necessary on bottomland sites which are too wet early in the spring (Wray and Schultz 1988).

One- or two-year-old bare-root seedlings can be planted using a tractor-pulled tree-planting machine (fig. 25), an auger (fig. 26), or a posthole digger or by hand with a shovel or "spud" (planting bar). With any planting technique, seedling root systems should not be exposed to wind or sunlight and should be kept moist in buckets of water or a thin mud-water slurry or covered with wet burlap. Seedlings should be planted about half an inch deeper than the old soil line on the seedling (Forbes 1971; Wray 1988). Spacing recommendations depend on the species and the approach used to establish the species. A list of possible spacing schemes and resulting number of trees per acre is given in table 4. Most seedlings can be planted at an 8-by-8-foot or 8-by-10-foot spacing (540 to 680 trees per acre). High-quality veneer species such as black walnut should be planted at 10-by-10-foot or 10-by-12-foot spacing (Wray 1988).

Tree-planting machines may be available for free or for a nominal rental fee from the Department of Natural Resources district forester (a list of district foresters and the counties included in their districts is provided in appendix 4), local county conservation board, or Soil Conservation Service district office (these are usually listed under County Government in the telephone directory). It is important that the planting machine be large enough to make a trench

Figure 25. A tree-planting machine on a three-point hitch. This planter has a 36-inch coulter, which is large enough to plant most one-year-old seedlings and some two-year-old stock. Photo by S. Pequignot.

Figure 26. A crew of six persons can efficiently hand-plant one- or two-year-old bare-root seedlings using a hand-held auger. Seedlings are kept in 5-gallon buckets of water until planted. Photo by R. C. Schultz.

Table 4. Number of Tree Seedlings Required for
Different Spacings

Spacing (Ft.)	Seedlings/Acre
4 × 4	2,722
5 × 5	1,742
6 × 6	1,210
6 × 8	907
7 × 7	889
6 × 10	726
8 × 8	680
10 × 10	436

that will accommodate seedling roots without forcing the seedling and bunching the root system. The machine's packing wheels should close and firm the trench around the seedling stem so that seedlings won't move upward when gently tugged. It may be necessary to have one person follow the planting machine on foot, tamping the soil around the seedlings to ensure that the trench is filled with soil. The planting machine can be used with a two- to four-person crew and is an efficient means of planting up to 800 trees per hour (Wray 1988). However, machine planting can only be done on sites accessible to tractors. In addition, the planting machine may impart an undesirable row appearance to the plantation. By mixing species within rows, creating irregular boundaries, planting at densities slightly higher than ultimately desired, and allowing natural mortality to thin the stand, it should be possible to simulate a more or less natural-looking canopy of overstory trees.

Auger- or hand-planting methods are more labor-intensive and may only be practical for small projects (2 to 3 acres), on sites inaccessible for large machinery, or for underplanting shrubs and understory trees after establishment of overstory species. Trees may be planted in any pattern (e.g., on centers or randomly spaced) using an auger or hand-planting techniques with a crew of four or more persons. Hand-held augers or tractor-mounted posthole diggers with at least an 8-inch-diameter bit are recommended to provide adequate space for seedling roots in the planting hole. Holes should be deep enough that the seedling can be planted half an inch deeper than the original soil line on the seedling without crowding the roots. With either the auger or hand-planting techniques, care should be taken to firm the soil around the seedling roots when it is replaced in the planting hole. Holes should be filled soon after they are made to prevent the soil from drying. Using a two-person hand-operated auger, four to eight persons can simultaneously drill holes and plant trees.

PLANTING UNDERSTORY SHRUBS AND HERBS. If understory tree and shrub species are planted at the same time as overstory species, careful monitoring of the project will be needed to ensure that vigorous growth of the understory species does not outcompete the species desired for the overstory (Tregay and Moffatt 1980). As a general rule, it is recommended that shrubs and herbs be added only after the canopy is established (Howell 1986). Successful introduction of understory plants has been done on a small scale using potted plants, bare-root seedlings, and seeds. Wade (1989) reports spreading topsoil borrowed from forest areas to introduce seeds and rhizomes of native species. Although it would take a long time to establish overstory species in this manner, this technique may be feasible for adding shrubs, herbs, and ground-layer vegetation after canopy establishment. The degree of successful establishment of understory plants varies widely and depends on the species involved and conditions (especially moisture availability) during the establishment period (Howell 1986).

SOURCES OF PROPAGULES

A number of commercial nurseries in Iowa sell planting stock which can be used in forest restoration. A partial list of these sources is given in appendix 2. In addition, Iowa Department of Natural Resources State Tree Nurseries located in Ames and Montrose sell seedlings (in quantity) of a number of tree and shrub species; however, there are statutory limitations on use of this stock. These nurseries and the stipulations pertaining to their stock are included in the appendix.

FOREST MAINTENANCE

The most critical initial maintenance on sites where no or few established trees are present is weed control. Pruning and training,

Figure 27. Herbicide was used to control weed growth in a 2-foot strip along this row of one-year-old green ash seedlings.

thinning, replacement plantings, and pest control may be necessary as well.

If desired overstory trees are planted as seedlings, weed control (particularly control of grasses such as smooth brome and fescue) is important during the first three to five years after planting (fig. 27), until the seedlings are tall or dense enough to suppress competition (Merritt 1980; Tregay and Moffatt 1980; Iowa Department of Natural Resources n.d.). Several methods may be used to limit weed growth, including mechanical cultivation, mowing, mulching, and chemical control.

Mechanical cultivation with a row cultivator, spring-tooth harrow, or other tractor-drawn device may be used on sites which are not prone to excessive erosion when soil is exposed. It is important to leave enough space between seedlings when they are planted to allow machinery in the plantation without damaging the trees. Maintain a distance of 6 to 12 inches between the blades and seedlings, and cultivate only to a depth of about 3 inches. It may be necessary to repeat cultivation as much as three to five times per year for the first five years (Wray 1988; Iowa Department of Natural Resources n.d.).

Mowing after seedlings are planted is probably the least desirable of the weed-control alternatives because it does little to decrease below-ground competition among plants on the site, and there is the possibility of damage to the seedlings. Mowing between rows of seedlings combined with herbicide treatment or mulching closer to the seedlings is recommended (Wray 1988; Iowa Department of Natural Resources n.d.). Mowing does serve to identify the location of the trees and is helpful in removing cover that otherwise might shelter rodents and rabbits that can girdle trees during winter.

Mulching is probably only practical for small-scale plantings. Sawdust, wood chips, bark, plastic, paper (hydromulch), or straw (if it can be removed before winter) will suppress competition in the immediate area of the seedling. Compost from municipal landfills is another potential mulching material and may be available at little or no cost.

Chemical treatment to control vegetative competition must be done by a certified individual using materials in accordance with label specifications. Appropriate herbicides are available from most distributors of agricultural supplies and are labeled for use with broad-leaved or coniferous trees. The herbicide to be used will depend on the type of vegetation (grasses, broad-leaved weeds, or woody plants) that must be controlled. Specific recommendations can be made by a Department of Natural Resources district forester, an extension forester, or a county conservation board biologist. Generally, a pre-emergent herbicide applied in early spring is recommended, followed by additional treatments when necessary (Iowa Department of Natural Resources n.d.; Wray 1988). Herbicide treatment is probably the most effective method for control of undesirable woody vegetation (some tree species are considered "weedy," for example, box elder or black locust); foliar application, stump application, or use of granular herbicides beneath unwanted trees or shrubs may be necessary (Iowa Department of Natural Resources n.d.).

In general, any weed-control technique eliminates cover and therefore usually limits damage to seedlings from rabbits and rodents. Depending on the site and climatic conditions, the presence of a few broad-leaved weeds may afford protection for tree seedlings from wind and water damage; removal of such weed growth is really only necessary if the seedlings are no longer exposed to sunlight.

Management for particular forest products, especially sawtimber from high-quality hardwoods, may require pruning and training as the trees reach sapling size. Timber-stand management practices and specific instructions for pruning, training, and thinning are given in Forbes (1971), Fazio (1985), Minckler (1975), Beattie et al. (1983), Wray (1989, 1990), and Wray and Farris (1989). Advice on managing habitat for various wildlife species is given in Bromley et al. (1990).

Even though there may be thousands of seedlings per acre in undisturbed forests, natural attrition results in only hundreds of saplings per acre and about a hundred mature trees per acre. If initial surveys of restoration areas indicate less than a 65 percent survival rate for planted seedlings, it may be necessary to do some replacement plantings so that the desired overstory species are able to capture the site (Forbes 1971; Wray 1988).

COST OF FOREST RESTORATION

There can be tremendous variation in the cost of forest restoration depending on the size of the area to be restored, how much site preparation is required, the number of trees to be planted per acre, and the type of planting stock used. The average cost for site preparation (mechanical or chemical) before tree planting was close to $75 per acre in 1991. Actual planting costs may vary between $150 (for one-year-old bare-root stock) to $1,000 (for larger container or balled-and-burlapped stock). Maintenance of the plantation for the first three to five years may cost close to $100 per acre per year, primarily for weed control. After establishment of the trees, maintenance costs should be minimal (Winebar and Gunter 1984; Dorney 1983; Twarok 1989; Wray 1989). Additional costs of $35 to $130 per acre may be expected if retiring the land from other uses (the equivalent of land rent for agricultural production). Direct seeding is a much less expensive option (Haynes and Moore 1988) but may be much less successful in establishment of overstory species, particularly for hardwoods. Cost sharing for tree establishment may be available from some government agencies (e.g., Agricultural Stabilization and Conservation Service, Soil Conservation Service, Conservation Reserve Program, Iowa Department of Natural Resources Resource Enhancement and Protection program, and the State Forest Reserve Act).

4 : Wetland Restoration in Iowa

HISTORY OF WETLANDS

THE MOST RECENT glaciation in north-central Iowa created a land-scape with numerous closed depressions on which prairies were interspersed with pothole marshes, shallow lakes, and a few deeper lakes. This prairie-wetland complex was the southern tip of a large area, including parts of Minnesota, North and South Dakota, Montana, Manitoba, Saskatchewan, and Alberta, known as the "prairie pothole region" (Kantrud and Stewart 1977). It is presently estimated that 2 to 3 million acres in north-central Iowa could have been wetlands prior to Euro-American settlement (Bishop 1989). In addition to these prairie pothole marshes, extensive wetlands existed on the floodplains and in backwater areas of the Mississippi and Missouri rivers and their major tributaries (Bishop and van der Valk 1982).

In 1850, the Federal Swamp Land Act granted 1,196,392 acres of public domain wetlands to the state of Iowa for "swamp reclamation" (Bishop 1981). This was followed by the establishment of drainage districts by the Iowa legislature, based on the premise that drainage of surface waters from agricultural and all other lands would be conducive to public health, convenience, and welfare (Bishop and van der Valk 1982). Drainage of wetlands in the state proceeded rapidly: USDA inventories indicated there were 930,000 acres of wetlands in 1906 and 368,000 acres in 1922, and Mann (1955) estimates that only 138,000 acres remained in 1955. Historical estimates are difficult to compare because there has been no generally agreed upon technical definition for wetlands until very recently (the revised "Federal Manual for Identifying and Delineating Jurisdictional Wetlands" was published in the *Federal Register* in July 1991 and is still in the review process). Although a few undisturbed or moderately disturbed wetlands remain, many of Iowa's current wetlands are probably severely degraded remnants of the state's original complex of pothole and riparian wetlands. The most recent estimates of remaining natural wetland areas in Iowa indicate there are only 26,470 acres of prairie pothole marshes and approximately 40,000 acres of oxbow and overflow wetlands associated with rivers and streams (Bishop 1981). These figures do not include a significant area of ephemeral and temporary wetlands or more specialized types of wetlands such as fens which are contained in Bishop and van der Valk's (1982) estimate of a total remaining wetland area of about 110,000 acres.

Along with drainage of inland "swamps," concomitant channelization of major rivers, especially the Missouri, contributed to the loss of large areas of critical riparian wetlands. Before Euro-American settlement, Iowa's major rivers followed meandering courses, naturally migrating across large floodplains and leaving oxbow lakes and overflow wetlands after flood events. However, as other wetlands were drained into the river systems and vegetation on the landscape was drastically altered, meandering rivers did not rapidly remove the additional runoff, resulting in severe flooding (Bishop 1981). Channelization allowed for more rapid removal of water and provided additional farmland in place of river bends and bottomland hardwoods. Extensive channelization along the Missouri River in the early 1900s eliminated nearly all natural wetlands along its course in Iowa and Missouri (e.g., Brinson et al. 1981; Baskett 1988). Along the Mississippi River, though, construction of a series of locks and dams created new wetlands, particularly in the northeastern corner of the state. Estimates of the pre-Euro-American settlement length of inland streams are more than double the 6,851 miles which remain, and the original area of associated riparian wetlands was undoubtedly significant although it is not specifically documented in the literature (Bishop 1981; Turner et al. 1981). The majority of Iowa's relatively undisturbed riparian zones are presently dominated by woody vegetation. Although most riparian wetlands have been lost to channelization and drainage, some have been dramatically altered in more recent years by inundation as a result of river impoundments and construction of reservoirs in the state.

TYPES OF WETLANDS IN IOWA

Wetlands are areas where surface water or groundwater flow patterns cause water to stay at or near the land surface for significant periods of time—for example, swamps, sloughs, marshes, potholes, bogs, seeps, natural lakes, rivers, river oxbows and overflow areas, and artificial wetland areas such as reservoirs, lakes, and ponds (Hubbard 1988). As such, water is a dominant factor determining the nature of soil development as well as the types of plant and animal communities living in the soil and on the surface. In a classification system for wetlands in the United States, Cowardin and coworkers (1979) define wetlands as:

> lands transitional between terrestrial and aquatic systems, where the water table is usually at or near the surface or the land is covered by shallow water. . . . Wetlands must have one or more of the following three attributes: (1) at least periodically, the land supports predominantly hydrophytes; (2) the substrate is predominantly undrained hydric soil; and (3) the substrate is non-soil and is saturated with water or covered by shallow water at some time during the growing season of each year.

Hydrophytic plants are those with special anatomical, morphological, and physiological features that allow them to survive and grow when partly or completely submerged by water. Frequently these plants are rooted in anaerobic soils (Bishop and van der Valk 1982). More than 470 aquatic and wetland plant species have been identified in Iowa (Lammers and van der Valk 1977, 1979). Only a brief description of dominant wetland species and community types

will be given in the following discussion of the three types of wetland systems found in Iowa. Distinguished on the basis of their
hydrologic regimes, these three include palustrine, lacustrine, and
riparian (riverine) systems (Bishop and van der Valk 1982; classification of Cowardin et al. 1979).

PALUSTRINE WETLANDS

Palustrine wetland systems are nontidal wetlands covering areas
less than 20 acres with water up to 6.6 feet deep (Cowardin et al.
1979). The prairie glacial marshes of north-central Iowa are shallow
basins within small watersheds that are classified as palustrine wetland systems (Bishop and van der Valk 1982). These wetlands have
seasonal fluctuations in water level reflecting annual rainfall patterns and annual changes in water level reflecting long-term climatic variations (drought cycles).

  Palustrine wetlands have been further separated into different
"types" based on more specific hydrological characteristics (Bishop
and van der Valk 1982; classification of Shaw and Fredine 1956).
Type I wetlands are "seasonally flooded basins" where soil is periodically waterlogged but drains well enough that it could be farmed
most years (fig. 29). These wetlands occur as upland depressions or
overflow bottomlands. The dominant vegetation depends on length
of inundation and may include smartweed (*Polygonum* spp.), wild
millet (*Echinochloa muricata*), fall panicum (*Panicum dichotomi-
florum*), sedges (*Carex* spp.), beggar-ticks (*Bidens comosa*), ragweed
(*Ambrosia* spp.), and barnyard grass (*Echinochloa crusgalli*) (Bishop
and van der Valk 1982). Type II wetlands are "freshwater meadows"
where soil remains waterlogged within inches of the surface and
standing water is present in spring and after heavy rainfall (fig. 30).
Species common in freshwater meadows include prairie cordgrass
(*Spartina pectinata*), reed canary grass (*Phalaris arundinacea*), common reed (*Phragmites australis*), manna grass (*Glyceria* spp.), sedges,
rushes (*Juncus* spp.), and mints (*Mentha* spp.). Type III wetlands (inland, shallow, freshwater) usually have at least 6 inches of standing
water, although they may dry out in late summer (fig. 31). Grasses,
bulrushes (*Scirpus* spp.), spikerushes (*Eleocharis* spp.), cattails (*Ty-
pha* spp.), arrowheads (*Sagittaria* spp.), giant bur reed (*Sparganium
eurycarpum*), smartweeds, and sedges are common in Type III wetlands. Type IV wetlands are "deep fresh marshes" with 6 inches to

Figure 29. A Type I palustrine wetland, Story County. Although it could be farmed in most years, in 1990 this area was too wet to plant.

Figure 30. A thicket of prairie cordgrass dominates a Type II palustrine wetland on Doolittle Prairie, Story County.

Figure 31. Normal water-level marks are noticeable on cattails surrounding the center of a Type III palustrine wetland basin in late summer, Dickinson County.

3 feet of standing water (fig. 32). Dominant vegetation in these wetlands includes cattails, bulrushes, reeds, spikerushes, bur reeds (*Sparganium* spp.), pondweeds (*Potamogeton* spp.), water milfoils (*Myriophyllum* spp.), coontails (*Ceratophyllum* spp.), waterlilies (*Nymphaea* spp.), and duckweeds (*Lemna* spp.).

The plant species in all four types of palustrine wetlands can be separated into communities according to the growth form of the dominant plants. Plant communities in Iowa wetlands include wet meadow communities composed of sedges, forbs, and grasses; mud-flat annual communities which include smartweeds and grasses; emergent communities made up of cattails, bulrushes, and reeds; floating-leaved communities which feature species such as water-lilies; free-floating communities which include duckweeds and watermeals (*Wolffia* spp.); and submersed communities which contain species such as pondweeds, coontails, and bladderworts (*Utricularia* spp.) (Bishop and van der Valk 1982). Most Iowa wetlands are composed of one or more of these plant communities, with gradual transitions between communities controlled primarily by water

Figure 32. Concentric zones of wetland vegetation communities are apparent bordering this Type IV wetland in Hamilton County.

depth. The typical sequence of plant communities in palustrine (and lacustrine) wetlands, going from shore to deeper water, is wet meadow→emergent→floating leaved→submersed (Bishop and van der Valk 1982). In a typical closed basin, this sequence appears as a series of concentric plant community zones centered around the deepest part of the basin (Kantrud et al. 1989). Free-floating plants often occur as an understory in the emergent community.

Seepage wetlands are a distinctive type of palustrine wetlands which form in areas of groundwater discharge (Bishop and van der Valk 1982). Typically, the soil is continuously saturated, although there may be little or no standing water. These wetlands are known as fens or hanging bogs (fig. 33). In some areas, the discharging groundwater may be very alkaline because it flows through calcareous substrates. In these situations, groundwater discharge creates a mix of precipitated carbonates in the soil which supports a number of plant species unique to this wetland type. Unique or rare species found in fens in Iowa include large arrow grass (*Triglochin maritima*), hooded ladies' tresses (*Spiranthes romanzoffiana*), yellow-

Figure 33. Fens support a unique assemblage of wetland plants.

lipped ladies' tresses (*Spiranthes lucida*), small white lady's slipper (*Cypripedium candidum*), small fringed gentian (*Gentianopsis procera*), Northern bog orchid (*Platanthera hyperborea*), and grass-of-Parnassus (*Parnassia glauca*), as well as some more common wet meadow and emergent community plant species (Roosa and Eilers 1978; Leoschke and Pearson 1988). Notable examples of alkaline fens in Iowa occur in the northwestern portion of the state, including Silver Lake Fen (Bishop and van der Valk 1982).

Palustrine wetlands in the prairie region are subject to changes in water level on a seasonal basis as well as long-term fluctuations due to natural drought cycles. Because of fluctuating water levels and muskrat activity, these wetlands have cyclical changes in vegetation, during which mudflat, emergent, or submersed and floating species replace each other as the dominant species (fig. 34). Four stages in the prairie marsh cycle are the dry marsh, regenerating marsh, degenerating marsh, and lake marsh (van der Valk and Davis 1978). The dry marsh condition develops during years of low water caused by below-normal precipitation, when all or a large part of the marsh may be nearly dry (at least have no standing water). Mudflat

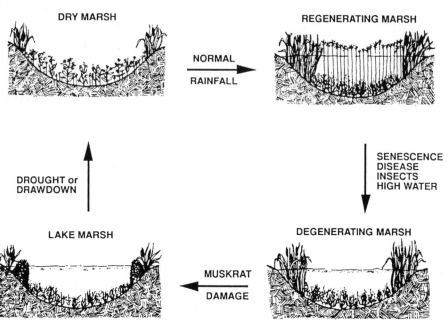

Figure 34. Vegetation changes in palustrine wetlands are caused by water-level fluctuations and changes in insect and animal populations. From van der Valk and Davis 1978, courtesy Ecological Society of America.

annual and emergent species are the dominant forms of vegetation. When normal precipitation resumes, standing water returns, allowing germination of submersed and free floating plants. This regenerating marsh stage is characterized by extensive cover of emergent species with submersed species as an understory. The next phase, the degenerating marsh, corresponds to the "hemi-marsh" stage of Weller and Spatcher (1965) and occurs when there is a rapid decline in populations of emergents due to a variety of interrelated factors which may include anoxia and damage from insects, disease, and muskrats. The hemi-marsh is characterized by a 25:75 to 75:25 ratio of emergent vegetation to open water and is the phase in which wildlife diversity and density are at a maximum (Kantrud et al. 1989). The subsequent lake marsh phase or "eat-out" is dominated by submersed and floating aquatic plant species, and the few remaining emergent plants may nearly disappear. The lake marsh phase continues until drought lowers the water level (which usually recurs on a five- to thirty-year cycle), returning the wetland to the dry marsh state (van der Valk and Davis 1978).

Figure 35. A lacustrine wetland in a protected bay of Little Wall Lake, Hamilton County. Willow trees provide a backdrop for arrowheads, cattails, reeds, and submersed vegetation.

LACUSTRINE WETLANDS

Lacustrine wetland systems are found in the shallow protected areas of lakes larger than 20 acres with water depth in the deepest part of the basin exceeding 6.6 feet. The wetlands in this system extend from the lakeshore to the point where at low-water stage the water is 6.6 feet deep (a more or less arbitrary point specified by Cowardin et al. 1979). These wetland areas typically are less subject to water-level fluctuations than those of the palustrine system. Lacustrine systems in Iowa are found along protected shores of Spirit Lake, Clear Lake, and many smaller lakes (Bishop and van der Valk 1982). Lacustrine wetland systems include the Type V wetlands of Shaw and Fredine (1956), also called "open freshwater" wetlands. Typical vegetation includes emergent species lining the shore, with floating and submersed communities in open water (fig. 35).

RIPARIAN WETLANDS

Although the classification of Cowardin and coworkers (1979) has been adopted for describing palustrine and lacustrine wetland systems, a broader definition of riparian systems will be used here. Their riverine wetland systems include only those wetlands contained within a river channel with flowing water that are not dominated by trees, shrubs, or persistent emergents, and they classify these areas as palustrine or lacustrine wetlands. In Iowa, very few strictly riverine wetlands occur, and these are found in backwater areas of the Mississippi and in the clearwater streams of the northeastern part of the state (Bishop and van der Valk 1982). Most Iowa streams and rivers are too turbid to support growth of aquatic plants because the plants are unable to get enough light or because they become buried by accumulating sediment.

The riparian wetlands as defined by Brinson et al. (1981) include the riverine system as well as overflow bottomland palustrine and lacustrine systems (e.g., oxbow lakes, backwater marshes) which occur in river floodplains and in their natural state become connected to the stream when it floods. This broader definition encompasses areas that function together as an ecosystem, where flowing water may supply moisture, sediments, and nutrients which would not be available in upland situations, regardless of the type of vegetation dominating the area. As such, riparian wetlands include the wooded bottomland areas along many of Iowa's rivers and streams. Species which may be present in overflow bottomlands include those described for palustrine and lacustrine plant communities as well as several species of bottomland trees and shrubs including willows (*Salix* spp.), cottonwood (*Populus deltoides*), silver maple (*Acer saccharinum*), green ash (*Fraxinus pennsylvanica*), box elder (*Acer negundo*), hackberry (*Celtis occidentalis*), slippery elm and American elm (*Ulmus rubra* and *U. americana*), basswood (*Tilia americana*), bur oak (*Quercus macrocarpa*), black walnut (*Juglans nigra*), poison ivy (*Rhus radicans*), honeysuckles (*Lonicera* spp.), dogwoods (*Cornus stolonifera* and *C. foemina*), gooseberry (*Ribes* spp.), raspberry (*Rubus* spp.), chokecherry (*Prunus virginiana*), elderberry (*Sambucus* spp.), and wahoo (*Euonymous atropurpureus*).

BENEFITS OF WETLAND RESTORATION

Benefits from the presence of wetland areas derive from these communities' role in maintaining environmental quality and protecting the welfare of wildlife populations and from the socioeconomic values resulting from direct and indirect human use (Tiner 1984; Maltby 1988; Leventhal 1990). Adamus and Stockwell (1983) have summarized the important functions of wetlands which can be related to the values just mentioned: groundwater recharge and discharge, flood storage and desynchronization, shoreline anchorage and erosion protection, sediment trapping, nutrient retention and removal, habitat for wildlife, habitat for fisheries, food chain support, and areas for passive and active recreation.

ENVIRONMENTAL QUALITY

Wetlands occupy a transitional position between terrestrial ecosystems and groundwater or surface water aquatic ecosystems. As such, wetlands often act as buffers or filters between urban development or agricultural systems and groundwater or surface water resources (Bastian and Benforado 1988). Although the role of wetlands in groundwater recharge alone is frequently presented as justification for their preservation, the relationships between groundwater and wetlands are complex and can change seasonally as well as spatially, with some basins acting as both groundwater recharge and discharge areas (Siegel 1988; Carter and Novitsky 1988). In any case, the close hydraulic contact between wetlands and groundwater points to the importance of wetlands in maintaining the quality of groundwater resources. Some of the hydrological functions of wetlands come simply by virtue of their physiographic position, for example, groundwater recharge and discharge functions and, to a certain extent, flood storage and flood peak moderation (Kittelson 1988). Other hydrological functions of wetlands depend on the presence of wetland vegetation, particularly erosion protection, sediment trapping, and nutrient removal and retention.

As Hubbard (1988) points out, the role of riparian wetland systems in flood attenuation and erosion control is well documented (e.g., Sather and Smith 1984), but the specific environmental quality functions of other wetlands, especially prairie glacial marshes, are just beginning to be understood. There is evidence that pothole marshes also play a significant role in water storage (hence flood desynchronization), but there is as yet little experimental verification for this (Hubbard 1988). It has been suggested that pothole wetlands in central Iowa provide enough water storage to be important to agricultural productivity. Modeling experiments predicted higher

surface soil moisture levels in an undrained watershed with potholes intact compared to a drained watershed (Campbell and Johnson
1975). Particularly in agricultural watersheds, both pothole wetlands and riparian wetlands remove and retain sediment, nutrients
(some in the form of fertilizers), and pesticides. Only a few studies
have been conducted on the role of prairie pothole wetlands in nutrient retention, but the available evidence indicates that, depending
on hydrologic conditions (drought vs. reflooding conditions), prairie
potholes can at least temporarily retain significant proportions of
influent nitrogen, phosphorus, and soluble organic carbon (Davis et
al. 1981). Merezheko (1973) has noted the importance of wetland
plants in filtration, absorption, and accumulation or volatilization
of these substances. However, as Kadlec (1988) indicates, the only
permanent mechanism for immobilization of substances entering a
closed depressional wetland is subsequent incorporation in the wetland soil. Because of their natural ability to purify water, some natural and created wetlands are being used as tertiary treatment areas
for sewage effluents (Bastian and Benforado 1988).

WILDLIFE HABITAT

The importance of Iowa's wetlands to many species of birds and
mammals cannot be overemphasized (fig. 36). The prairie-wetland
complexes of the prairie pothole region provide the most important
breeding and nesting sites for the North American population of
dabbling ducks such as the mallard, green- and blue-winged teal,
northern pintail, and northern shoveler (Hubbard 1988). Other frequent nesters include the redhead, ruddy duck, and wood duck. Both
pothole wetlands and riparian wetlands in Iowa are essential to
countless other migratory birds including the Canada goose, snow
goose, and bald eagle that frequent the Missouri and Mississippi fly-
ways (Dinsmore et al. 1984). Since the time of Euro-American settlement, several species of aquatic birds have been extirpated as nesters
in Iowa: sandhill crane, whooping crane, common loon, trumpeter
swan, swallow-tailed kite, piping plover, and least tern (the last two,
however, have nested in Iowa along the Missouri River in recent
years). Although some of these species (the first four listed) were
hunted, direct habitat loss was probably the primary factor in their
disappearance from the state. Additional aquatic birds whose future
as breeding species in Iowa might be in jeopardy include the eared

Figure 36. Canada geese congregate on the shores of Hendrickson Marsh, Story County.

grebe, great blue heron, great egret, black-crowned night heron, American bittern, bald eagle, king rail, and Forster's tern (Dinsmore 1981). Several wetland bird species seem to nest only in larger wetlands (50 to 75 acres) or in wetland complexes (Brown and Dinsmore 1986). Other birds which still commonly appear in Iowa wetlands include yellow-headed and red-winged blackbirds, pied-billed grebe, American coot, green-backed heron, spotted sandpiper, and common yellowthroat (Dinsmore 1981; LaGrange and Dinsmore 1989; Dinsmore et al. 1984). Remaining forested riparian wetlands provide habitat for several species of woodpeckers, owls, hawks, vireos, and warblers. Dinsmore (1981) states that some species which utilize forested riparian habitats may be threatened by fragmentation and loss of habitat, for example, the red-shouldered hawk. Both upland depressional and riparian wetlands can provide important winter cover for common winter resident game-bird species such as pheasant and partridge.

Although considerable research effort has been directed toward gaining an understanding of the relationships between waterfowl and wetlands, much less is documented concerning the relationships between mammals and wetlands. As Fritzell (1988) points out, only a few mammals are adapted for living solely in wetland environments. However, wetlands may provide important resources for a number of mammalian species that occupy rather broad ecological niches. Wetlands supply habitat and/or prey for the muskrat, otter, beaver, mink, striped skunk, long-tailed weasel, raccoon, and red fox, and they provide important winter cover for white-tailed deer (Hubbard 1988). The important ecological role of muskrats in the cyclical vegetation dynamics of pothole marshes has already been mentioned. Other mammals commonly associated with wetlands include mice, voles, and shrews. A larger array of scansorial and herbivorous mammals may be common to bottomland hardwood wetlands (Fritzell 1988). The distribution of riparian wetlands as corridors across Iowa's largely agricultural landscape provides protected pathways for animal movements and migrations (Brinson et al. 1981).

Degradation of aquatic environments and loss of habitat due to agricultural and urban development have also had a significant negative influence on populations of amphibians, reptiles, and fishes in Iowa. Many species of frogs, toads, lizards, salamanders, turtles, and snakes that occur in the state are wetland dependent. Iowa's present population includes roughly seventy-six species and subspecies from these groups (Christiansen 1981). Among fishes, several species of minnow, shiner, catfish, bass, perch, and some sunfish, pike, trout, and sticklebacks appear in Iowa's inland waters and border rivers (Menzel 1981).

The abundance and diversity of aquatic invertebrates found in wetlands provide a critical link in the detrital-based food webs of these communities (Murkin and Wrubleski 1988; Fritzell 1988). Many wetland animals, including fish, amphibians, reptiles, birds, and a number of mammals, are insectivores. Invertebrates are noted as a valuable high-protein food source for breeding birds and juvenile waterfowl (Hubbard 1988).

SOCIAL VALUES

Socioeconomic benefits derived from wetlands include those from direct and indirect human use of these areas. Hunting, trapping, fishing, hiking, wildlife observation, and educational experiences are dominant forms of active recreational use of Iowa wetlands.

Other, less direct benefits include heritage values, such as preservation of endangered or rare plants and animals, many of which are wetland dependent (Neiring 1988).

WETLAND ESTABLISHMENT

Natural wetlands are extremely complex systems interacting with both local and regional water tables and are comprised of numerous species which have coevolved over a relatively long period of time (Hook 1988; Zedler 1988). In addition, wetlands are dynamic ecosystems in which successional and/or cyclical changes in dominant plant and animal species occur (van der Valk and Davis 1978; van der Valk 1981, 1985; Odum 1988). The specific links between the physical, biological, and hydrological characteristics of wetland ecosystems and their roles in performing water-quality enhancement and providing wildlife habitat are not yet fully understood. Some have suggested that, particularly for prairie pothole wetlands, the hydrological regime is a primary factor in determining performance of wetland functions (Larson 1988; Hubbard 1988). As Fischel (1988) indicates, a properly restored wetland may require several years before beginning to function as a "natural" wetland. Hydrological functions can usually be restored fairly rapidly, while nutrient accumulation functions take longer to achieve, and the ability of the wetland to support diverse wildlife populations may take a number of years to develop.

In order to simulate natural wetland biology (and thereby attempt to simulate natural wetland functions) in wetland restoration, careful consideration must be given to site selection (suitability of soils and landforms for growth of desired species), site preparation (feasibility of creating the desired hydroperiod through use of natural runoff patterns, interruption of existing drainage structures, and/or installation of water control devices), methods of establishing wetland vegetation, and methods for maintenance of the wetland after it is established.

SITE SELECTION

Former wetland sites as determined by the presence of hydric soils have the greatest potential for successful wetland restoration (Iowa Department of Natural Resources 1988). A hydric soil is "a soil that in its natural condition is saturated, flooded or ponded long enough during the growing season to develop anaerobic conditions that favor the growth and regeneration of hydrophytic vegetation" (USDA 1987).

The presence of water on or in soil for extended periods of time results in chemical and physical changes that lead to the development of certain soil morphological features which are recognizable for many years after the soil has been artificially drained (Buol and Rebertus 1988; Parker 1988). Distinctive hydromorphic soil features which reflect extended saturation include predominantly olive and/or gray colors (gley colors) and accumulation of organic matter (sometimes to the point of formation of organic soil horizons) at the soil surface. Variable or mottled soil color patterns of brown or yellow spots in a gray or olive-gray matrix also indicate relatively long periods of saturation (Parker 1988). The Iowa Soil Properties and Interpretations Database (ISPAID 1989) identifies soil map units used in county soil surveys throughout the state that consistently meet the definition of hydric soils. The county soil surveys (which are available at county Soil Conservation Service offices or at the Soil Survey Laboratory, Iowa State University) contain detailed soil maps which can be used to identify potential restoration sites. A reasonably comprehensive list of soil series names for hydric soils with potential for wetland restoration was gathered from this data base and appears in table 5. The soil map units used in soil surveys are phases of soil series based on features that affect soil use and management (e.g., percent slope). As such, only specific phases of the series listed in table 5 may be appropriate for wetland restoration. Descriptions of soil map units are given in the soil survey and should be used to determine the feasibility of wetland restoration only on a site by site basis. At the scale of Iowa's county soil survey maps, the minimum delineation of a map unit is 2 acres, so some small basins suitable for restoration may be indicated only by a symbol for a marsh or "wet spot."

Since nearly all of Iowa's natural wetlands have been drained, with concomitant lowering of water tables, the potential for maintaining restored wetlands on sites with hydric soils may be somewhat limited. The landscape position and size of the watershed, soil permeability, depth to water table, and direction of net movement of groundwater must be carefully evaluated on a site by site basis to

Table 5. Hydric Soils with Potential for Wetland Restoration (by Series Name)

Upland/ Depressional Basins	Riparian Zones	Upland/ Depressional Basins	Riparian Zones
Adrian	Ackmore	Letri	McPaul
Ames	Afton	Madelia	Millington
Ansgar	Albaton	Marcus	Napa
Appanoose	Amana	Marna	Nishna
Beckwith	Ambraw	Maxfield	Nodaway
Belinda	Blend	Minnetonka	Okaw
Blue Earth	Bremer	Muskego	Ossian
Boots	Buckney	Niota	Otter
Brownton	Calco	Okoboji	Owego
Calcousta	Caneek	Palms	Raccoon
Canisteo	Carlow	Revere	Radford
Clyde	Chaseburg	Rolfe	Rowley
Cordova	Chequest	Rubio	Sawmill
Corley	Coland	Rushville	Shandep
Darfur	Colo	Sperry	Snider
Denrock	Coppock	Spicer	Solomon
Edina	Darwin	Taintor	Spillville
Faxon	Dolbee	Wacousta	Terril
Garwin	Dorchester	Waldorf	Tilfer
Haig	Elvers	Webster	Titus
Harcot	Elvira	Winterset	Tripoli
Harps	Forney	Zwingle	Tuskeego
Harpster	Hanlon	"marsh"	Wabash
Houghton	Havelock	"muck"	Woodbury
Jameston	Holly Springs		Zook
Kalona	Humeston		"Aquents"
Knoke	Huntsville		"Fluvaquents"
Kossuth	Lawson		"riverwash"
Lanyon	Luton		

Source: ISPAID 1989.

determine the feasibility of creating the desired water regime (Larson 1988). The size of the wetland restoration project will be determined by local conditions as well as by economic constraints. The possibility of saturating or flooding nearby land areas, particularly those under different ownership, must also be considered (Iowa Department of Natural Resources 1988). Wetland restoration should not be undertaken without consulting local drainage district officials.

If wetland plants are to be established by natural regeneration of local propagules (other options are discussed in a subsequent section), length of time since drainage, how frequently ponding might have occurred since drainage, frequency of herbicide use, and possibly proximity to other wetland areas will largely determine species diversity and should be considered during site selection. Seeds of some prairie pothole wetland species may be present in the wetland seed bank for several decades after drainage due to seed longevity and periodic recruitment (van der Valk and Davis 1978; Erlandson 1987). LaGrange and Dinsmore (1989) report an average of twenty-four plant species per wetland on three restored marshes in central Iowa that had been drained for at least forty years. Weinhold and van der Valk (1989) note reduced species diversity and reduced seed density in seed banks of pothole marshes that had been drained for five or more years. According to Weinhold and van der Valk (1989), seed banks of wetlands drained for fewer than twenty years contain viable seeds of many wetland species and may be among the best candidates for restoration. However, many of Iowa's former wetland areas have been drained for over seventy years. An inspection of the "weeds" present in a potential restoration area or an examination of the seed bank of a drained wetland might be useful before restoration efforts are begun (Pederson and Smith 1988). Wind-dispersed seeds of a number of species from nearby wetlands or seed dispersal by wetland-dwelling wildlife (particularly ducks) may also contribute to revegetation of restored wetlands. Proximity to other wetlands may also be important in order to establish a viable muskrat population, which may be necessary for maintaining the usual cycle of vegetation in prairie pothole wetlands.

SITE PREPARATION

During years of normal or above normal precipitation, many successful restorations of drained pothole wetlands have been accomplished simply by interruption of existing drainage facilities, for ex-

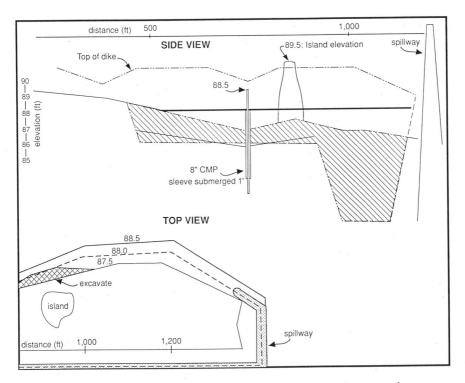

Figure 37. Plan for construction of a simple pond and spillway according to Soil Conservation Service specifications. Courtesy of J. Swartzendruber.

ample, excavating or plugging drainage tile lines on the periphery of the wetland basin or installing an unperforated standpipe directly on the tile line within the basin and then utilizing normal runoff patterns to pond water (LaGrange and Dinsmore 1989; Madsen 1988). Construction of earthen dams across open ditches is also commonly used to impede drainage. Due to the accretion of sediment and lower groundwater levels, excavation of the central portion of the pothole using dynamite or large earth-moving equipment may provide for larger and deeper open-water areas (LaGrange and Dinsmore 1989) and make available soil material for construction of low-head dikes (fig. 37). Construction specifications for dikes vary among different agencies. A height of 2 to 4 feet and top width of 6 to 10 feet with 2:1 to 3:1 side slopes would be considered adequate in most small watersheds (USDA 1984; Fischer 1990). Soil Conservation Service specifications for wetlands suggest a goal for water depth of 3.5 feet in at least half the basin (USDA 1984), although many wetlands have been designed for as little as 1.5 to 2 feet of water. Installation of devices to control water levels can be useful in management of pothole water regimes to favor certain plant and animal species (Kantrud 1986).

Incentives for wetland restoration projects have been initiated by different government agencies, including cost-share programs for construction and inclusion of wet areas in set-aside acreage where government personnel will be allowed to install drainage impediments (consult local Soil Conservation Service or U.S. Fish and Wildlife Service personnel). In most cases, restoration projects supported by government funds are subject to approval by the specific agency involved and may require meeting particular design specifications depending on the size of the watershed. Publications containing construction information are currently being prepared by the U.S. Fish and Wildlife Service and the Soil Conservation Service.

Site preparation for restoration of riparian wetlands may entail interruption of drainage tiles as well as removal of levees, dikes, or other river-impoundment structures. Some excavation of floodplains may be necessary to create ponding sites (Hey 1988). For establishing bottomland hardwoods, mechanical or chemical removal of weeds, especially grasses, and undesirable woody vegetation is recommended before planting. If heavy sod cover is present, tillage or chemical treatment of 4-foot-wide strips where trees are to be planted may be necessary (see chapter 3).

METHODS OF ESTABLISHING WETLAND VEGETATION

To date, many restorations of prairie pothole marshes in Iowa have relied almost exclusively on natural regeneration by local propagules (Heiser 1989). Wetland vegetation usually becomes established within a year of creating ponded conditions on former wetland sites (Weinhold and van der Valk 1989; Heiser 1989). However, since many of Iowa's wetlands have been drained in the early 1900s, it may be desirable to augment the native seed bank by seeding in locally collected seed from extant wetlands or by inoculating the site with sediments containing a more diverse seed bank (Pederson and Smith 1989; Larson 1988). Seeding or inoculating the site is mechanically easier to accomplish during a drawdown, but exact timing may depend on seed dormancy and germination requirements. Wetland restorationists active in central Iowa are not aware of any currently available commercial seed sources for local (Iowa) genotypes of palustrine wetland species (Lekwa 1989; Heiser 1989). A more thorough discussion of the importance of local genotypes in native community reconstruction was given earlier with respect to prairie restoration. There are firms in neighboring states that sell wetland plant seed and wetland plants which may be appropriate for parts of Iowa near the seed source (primarily in Minnesota and Wisconsin). These are included in appendix 3.

For establishment of bottomland hardwoods, natural regeneration of light-seeded wind- or water-disseminated species such as cottonwood, willow, and elm can be expected if local populations are present in the area. Direct seeding and use of seedlings from nursery stock or adjacent natural stands have been successful methods for establishing hardwoods on abandoned agricultural bottomlands in the southeastern United States (Haynes and Moore 1988). Sources of Iowa propagules are given in appendix 2. Again, many species of bottomland hardwoods (especially green ash, silver maple, and the oaks) in the central states are most successfully established when planted as seedlings. To achieve optimal species diversity, it may be desirable to seed in or plant shrub seedlings as well. Native shrubs are also available from many of the sources listed in appendix 2. Herbaceous understory species may naturally invade after the canopy is established or may be added as seeds or seedlings at that time (Howell 1986).

Spring (April 1 to May 15) is the best time for planting bare-root tree seedlings in Iowa, but fall planting may be done if bottomland sites are too wet in spring (Wray and Schultz 1988). More detailed information for establishment of trees and shrubs is given in chapter 3.

If preservation of palustrine riparian wetlands improves water quality significantly, natural restoration of riverine wetlands will likely occur without intervention. Otherwise, restoration of strictly riverine wetlands is probably not feasible.

WETLAND MAINTENANCE

POTHOLE WETLANDS

Species composition of natural wetland vegetation communities depends on the species present in the seed bank, the water regime, and disturbances to the ecosystem (Hubbard 1988). Natural disturbances in the form of drawdowns due to variations in climatic conditions of the prairie are an important component in the successional cycle of prairie pothole wetlands (van der Valk and Davis 1978; Hubbard 1988). Light grazing and occasional fires were probably also historically important in controlling species composition in natural prairie wetlands (Kantrud 1986). On restored but unmanaged wetlands in the prairie pothole region, successful establishment of species considered valuable for wildlife, such as bulrushes, may be followed by invasion and development of monotypic stands of cattails or common reed which are less attractive to a diversity of wildlife species (fig. 38). Artificial drawdowns, prescribed burns, and limited mowing or very light grazing might be used on restored pothole wetlands to mimic natural disturbances which control normal successional patterns in these marshes (Odum 1988). It may be necessary to manage (i.e., burn or graze) only a portion of a pothole in order to leave undisturbed refuge areas for reptiles and amphibians.

Although it is possible to maintain a natural successional pattern through management of restored marshes, so far it has not been possible to maintain the ideal hemi-marsh stage for an extended period of time (Lekwa 1989). A viable population of muskrats seems necessary to continue the normal prairie marsh vegetation cycle, so maintenance of a minimum residual water area even during drawdown events is required if there is no nearby source of muskrats to repopulate the restored wetland (van der Valk and Davis 1978; Lekwa 1989). Installation of water-level control structures to regulate artificial drawdowns may not be economically feasible for a large number of pothole restorations. It may be more realistic to rely on natural climatic variation to regulate water levels as much as possible.

Reviewing the scarce literature on the effect of fire in prairie glacial marshes, Kantrud (1986) cautions that not enough is known to formulate prescriptions for burning. Fire has been observed to be helpful in controlling dense stands of emergent species such as cattails and common reed. Only a portion of the wetland area should be burned at any one time. Kantrud emphasizes that in the absence of water-level control, fall burning to eliminate emergent cover of seasonally ponded basins reduces their ability to trap snow and may shorten the period of standing water in the following year.

Kantrud also summarizes the effects of grazing on prairie pothole wetland plant communities, indicating that periodic light grazing

Figure 38. An unmanaged wetland may become a single-species cattail marsh like this one in Dickinson County.

may increase species diversity and create more intricate patterns among plant communities. However, in some wetland types, especially fens, grazing can destroy the properties of the substrate and result in serious damage to plant and animal populations.

RIPARIAN WETLANDS

Control of competing vegetation, especially grasses, is important in establishing tree seedlings in the Midwest (Wray 1988). Possible methods of weed control are discussed in some detail in chapter 3. If chemical weed control is used, it is vital that herbicides be applied with extreme caution in riparian ecosystems. Spot or strip treatment to remove all competing vegetation in bands 3 to 4 feet wide around the seedlings is recommended until the canopy of the trees is sufficient to eliminate competition.

COST OF WETLAND RESTORATION

Little information is available concerning the cost of wetland restoration. Data compiled by Heiser (1989) and by Fischel (1988) indicate that the cost of relatively simple pothole wetland restoration may average between $150 and $250 per acre (for earth moving and installation of tile plugs). Projects requiring installation of sophisticated water-control structures may incur costs of thousands of dollars per acre (Hey 1988). In 1991, the cost for site preparation, establishment, and maintenance of bottomland hardwoods was just over $200 per acre, assuming low-cost seedlings are used. Additional costs of $35 to $130 per acre may be expected to cover the cost of retiring the land from other uses (the equivalent of land rent for agricultural production or pasture use; the lower values are common in the southern part of the state) (Twarok 1989; Madsen 1988).

Epilogue

TOO OFTEN human activities have led to the featureless monotony of single-species cultures or pavements over the landscape. Now, more than ever before, there is widespread recognition that some of our tinkering with nature should include the restoration of indigenous flora and fauna in their rightful places. I hope this book will serve as an inspiration to those who own or manage portions of the Iowa landscape, however small or large the area may be. It is possible, given a bit of time, some financial resources, and a lot of patience, to reconstruct a part of our "beautiful land."

Iowa's natural landscape communities can speak for themselves more eloquently than anyone can speak for them. Everyone can derive inspiration from plunging into a 200-acre (or larger) prairie, wandering along the riverbank in a streamside forest, or listening to the honking of geese on the edge of a marsh. Certainly a visit to one of our state preserves would be requisite for the beginning restorationist.

Most natural-area restoration has been done on an experimental basis. Continue the experimentation! Although this book is meant to be a fairly thorough summary of restoration techniques that have met with success, the possibilities are probably endless. Don't be afraid to try something new and different, although it would be best to conduct your more daring experiments at a relatively small scale. In any case, you will need to be creative to tailor the ideas presented herein to your own site.

Appendix 1.

PRAIRIE SEED AND PLANT SOURCES

Albert Lea Seed. P.O. Box 127, Albert Lea, MN 56007. (507) 373-3161.

Dan Allen. R.R. 2, Box 311, Winterset, IA 50273. (515) 462-1241.

Arrow Seed Co. (forbs). P.O. Box 722, Broken Bow, NE 68822.

Bluestem Seed Company (forbs). Highway 46 East, Grant City, MO 64456. (816) 786-2401.

Jim Burdette. R.R. 2, Diagonal, IA 50845. (515) 734-5793.

Howard Christiansen. R.R. 1, Box 20, Wiota, IA 50274. (712) 783-4230.

Kathy Clinebell (forbs). R.R. 2, Box 176, Wyoming, IL 61491. (309) 286-7356.

County Road Greenhouses (forbs). R.R. 1, Box 74, Malta, IL 60150.

John Curry. R.R. 2, Box 101, Exira, IA 50076. (712) 549-2216.

Earl May Seed and Nursery. 208 N. Elm, Shenandoah, IA 51601. (712) 246-1020.

E. F. Mangeldorg Seed (forbs). 4500 Swan Ave., P.O. Box 327, St. Louis, MO 63166. (314) 535-6700.

Farmers Co-op. 300 Osage St., Creston, IA 50801. (515) 782-6411.

Lee Farris. R.R. 1, Box 75, Mount Ayr, IA 50854. (515) 464-3671.

Henry Field Seed and Nursery. 407 Sycamore, Shenandoah, IA 51602.

Horizon Seed Co. (forbs). 1600 Cornhusker Highway, P.O. Box 81823, Lincoln, NE 68501. (402) 475-1232.

Iowa Prairie Seed Co. (forbs). 110 Middle Rd., Muscatine, IA 52761. (319) 264-0562.

Kester's Wild Game Food Nurseries (forbs). 4488 Highway 116 East, P.O. Box V, Omro, WI 54963.

LaFayette Home Nursery (forbs). R.R. 1, Box 1A, LaFayette, IL 61449. (309) 995-3311.

Little Valley Farm (forbs). R.R. 1, Box 278, Richland Center, WI 53581.

L. L. Olds Seed Co. (forbs). P.O. Box 7790, Madison, WI 53707. (608) 249-9291.

Keith McGinnis. 309 E. Florence, Glenwood, IA 51534. (712) 527-4308.

Midwest Wildflowers (forbs). P.O. Box 64, Rockton, IL 61072.

Missouri Wildflowers Nursery (forbs). 9815 Pleasant Hill Rd., R.R. 2, Box 373, Jefferson City, MO 65109.

Natural Garden (forbs). 38 W. 443 Highway 64, St. Charles, IL 60174.

Nature's Way (forbs). R.R. 1, Box 62, Woodburn, IA 50275. (515) 342-6246.

Olympic Seed Co. P.O. Box 752, Highway 248 South, Independence, IA 50644. (319) 334-7373.

Duane Ortgies. 905 Birch St., Atlantic, IA 50022. (712) 243-3814.

John Osenbaugh (forbs). R.R. 1, Box 106, Lucas, IA 50151. (515) 766-6792.

Ottilie Seed Farms. R.R. 1, Highway 14 North, Marshalltown, IA 50158. (800) 542-7894.

Prairie Bluestem Nursery (forbs). R.R. 2, Box 92, Hillboro, IL 62049.

Prairie Moon Nursery (forbs). R.R. 3, Box 163, Winona, MN 55987.

Prairie Nursery (forbs). P.O. Box 365, Westfield, WI 53964. (608) 296-3679.

Prairie Restorations (forbs). P.O. Box 327, Princeton, MN 55371.

Prairie Ridge Nursery (forbs). R.R. 2, 9738 Overland Rd., Mt. Horeb, WI 53572. (608) 437-5245.

Prairie Seed Source (forbs). P.O. Box 83, North Lake, WI 53064.

Richard Routh. 406 W. Washington, Mount Ayr, IA 50854. (515) 464-2240.

Sexauer Co. 444 S.W. Fifth, Des Moines, IA 50309. (515) 288-0238.

Sharp Bros. Seed Co. (forbs). P.O. Box 140, Healy, KS 67850. (316) 398-2231.

Sharp Bros. Seed Co. (forbs). P.O. Box 665, Clinton, MO 66735.

Stock Seed Farms. R.R. 1, Box 112, Murdock, NE 68407. (402) 867-3771.

Wayne Vassar A-G Grain and Seed. R.R. 4, Clinton, MO 64735. (816) 885-8521.

Wehr Nature Center (mixed forbs). 9701 W. College Ave., Franklin, WI 53132. (414) 425-8550.

Wildlife Nurseries (forbs). P.O. Box 2724, Oshkosh, WI 54903. (414) 231-3780.

Appendix 2.

TREE NURSERIES AND SEED DEALERS IN IOWA

Adams Nursery. 203 E. Ninth, Monticello, IA 52310. (319) 465-3626. Balled-and-burlapped conifers.

Ada's Flowers–Nursery–Antiques. 1107 First St., Keosauqua, IA 52565. (319) 293-3118. Balled-and-burlapped and container conifers and hardwoods.

Allen Garden Center. 5491 S.E. Sixth Ave., Des Moines, IA 50317. (515) 266-7170. Bare-root conifers.

Arns Nursery. R.R. 2, Waverly, IA 50677. (319) 885-6420. Balled-and-burlapped conifers and hardwoods.

Ashton Tree Farm. R.R. 1, Ashton, IA 51232. (712) 724-6343. Container conifers and hardwoods.

Ashworth Acres Tree Farm. R.R. 10, West Des Moines, IA 50265. (515) 225-2654. Tree-spade conifers and hardwoods.

Bennett Landscape Co. P.O. Box 16, State Orchard Rd., Council Bluffs, IA 51501. (712) 323-3356. Balled-and-burlapped and container conifers and hardwoods.

Blackmore Nursery. 650 S. Garfield, Mason City, IA 50401. (515) 424-1213. Balled-and-burlapped, container, and tree-spade conifers and hardwoods.

Bluebird Hill Nursery Service. R.R., Sperry, IA 52650. (319) 985-2415. Bare-root stock in quantity, also larger conifers and hardwoods.

Boone River Tree Farm. 301 Cedar Ave., Eagle Grove, IA 50533. (515) 448-4418. Tree-spade conifers and hardwoods.

Boyd Nursery. R.R. 3, New Hampton, IA 50659. (515) 394-4745. All types of stock.

Brewer's Crescent Nurseries. 700 Old Lincoln Hwy., Crescent, IA 51526. (712) 545-3265. Balled-and-burlapped, container, and tree-spade conifers and hardwoods.

Burton's Trees. Battle Creek, IA 51006. (712) 365-4489. Bare-root and container conifers.

Cascade Forestry Service. R.R. 1, Cascade, IA 52033. (319) 852-3042. Bare-root, balled-and-burlapped, and container conifers and hardwoods.

Central Iowa Landscaping Contractors. R.R. 2, Box 187, Waukee, IA 50263.

(515) 987-1609. Bare-root, balled-and-burlapped, and container conifers and hardwoods.

C. K. Nursery and Landscaping. R.R. 2, Box 154, Woodard, IA 50276. (515) 438-2814. Bare-root, balled-and-burlapped, and container conifers and hardwoods.

Clarinda Windbreak Tree Co. R.R. 1, Clarinda, IA 51632. (712) 542-5671. Bare-root, balled-and-burlapped, and container conifers.

Correctionville Nursery. R.R. 1, Correctionville, IA 51016. (712) 372-4485. Bare-root, balled-and-burlapped, and container conifers; balled-and-burlapped, container, and tree-spade hardwoods.

Country Garden Store. 1639 W. James St., Webster City, IA 50595. (515) 832-3173.

Country Landscapes. R.R. 2, Ames, IA 50010. (515) 232-6864. Balled-and-burlapped and container conifers and hardwoods.

Cousins Nursery. P.O. Box 304, Grafton, IA 50440. (515) 748-2316. Container conifers.

Danamere Nursery. R.R. 1, Carlisle, IA 50047. (515) 989-0229. Balled-and-burlapped, container, and tree-spade conifers and hardwoods.

Del's Garden Center. Hwy. 13E, Spencer, IA 51301. (712) 262-4912. All types of stock except container conifers.

Don's Planting and Landscaping. 934 Lisa Dr., Waterloo, IA 50701. (515) 234-4052 or (515) 277-2060. Bare-root conifers; balled-and-burlapped, container, and tree-spade conifers and hardwoods.

Dufford. R.R. 1, Shell Rock, IA 50670. (319) 885-6234. Hardwood seedlings in quantity.

Dunn Tree Farm. R.R. 1, Box 54, Adel, IA 50003. (515) 993-4960. Balled-and-burlapped conifers and hardwoods; container and tree-spade conifers.

Earl May Seed and Nursery. 208 N. Elm, Shenandoah, IA 51601. (712) 246-1020. All types except tree-spade stock.

Ekstam Tree Farm. R.R. 2, Box 260, Laurens, IA 50554. (712) 845-2013. Container and tree-spade conifers and hardwoods.

Feldman Bros. Landscaping. R.R. 1, Box 262, Letts, IA 52754. (319) 726-4222. Hardwood seedlings in quantity; balled-and-burlapped and container hardwoods.

Ferris Nursery. 811 Fourth St. N.E., Hampton, IA 50441. (515) 456-2563. Seedlings in quantity; balled-and-burlapped and container conifers and hardwoods.

Fitzpatrick's Nursery. P.O. Box 68, Lynville, IA 50153. (515) 527-2430. Hardwood and conifer seedlings in quantity; balled-and-burlapped and tree-spade conifers and hardwoods.

Forrest Burton's Evergreen Sales. 1103 Eighth St., Grand Junction, IA 50107. (515) 738-2605. Conifer seedlings in quantity; balled-and-burlapped, container, and tree-spade conifers.

Fort Atkinson Nursery. Fort Atkinson, IA 52144. (319) 534-7273. Hardwood seedlings in quantity; bare-root, balled-and-burlapped, and container conifers and hardwoods.

Fuggiti Nursery. 530 E. Eagle, Waterloo, IA 50701. (319) 342-3220. All types of stock.

Green Action. R.R. 3, Waverly, IA 50677. (319) 987-2224. Hardwood seedlings in quantity; container hardwoods.

Greenery. Airport, Elkader, IA 52043. (319) 245-1458. Bare-root conifers in quantity.

Green Thumbers. 5112 Grandview Ave., Muscatine, IA 52761. (319) 263-4403. Hardwood seedlings in quantity; balled-and-burlapped and tree-spade hardwoods.

Greenworld. 309 Seventh St. N.W., Sioux Center, IA 51250. (712) 722-2621. All types of stock.

Grupe Nursery. 2735 Mt. Pleasant St., Burlington, IA 52601. (319) 752-2135. Balled-and-burlapped and container conifers and hardwoods.

H & M Industries. (c/o M. Dusenberg). P.O. Box 252, Clear Lake, IA 50428. (515) 357-2825. Bare-root conifers and hardwoods.

Heard Gardens. 5355 Merle Hay Rd., Des Moines, IA 50323. (515) 276-4533. Balled-and-burlapped, container, and tree-spade conifers and hardwoods.

Henry Field Seed and Nursery. 407 Sycamore, Shenandoah, IA 51602. Seed for nut trees.

History Hill Tree Farm. R.R., Randolph, IA 51649. (712) 625-2791. Balled-and-burlapped and tree-spade conifers and hardwoods.

Hoesing Tree Farm and Nursery. Mapleton, IA 51034. (712) 882-2741. All types of stock.

Hoffman Tree Farm. 9409 L Ave., Marion, IA 52302. (319) 377-0977. Balled-and-burlapped and tree-spade conifers; tree-spade hardwoods.

Holub's Country Nursery and Greenhouse. 429 E. Eighth St., Cresco, IA 52136. (319) 547-3521. R.R. 2, Ames, IA 50010. (515) 232-4769. All types of stock.

Horak's Landscape Nursery. R.R. 1, Swisher, IA 52338. (319) 848-7222. Balled-and-burlapped, container, and tree-spade conifers and hardwoods.

Howell Tree Farm. Cumming, IA 50061. (515) 981-4762. Tree-spade conifers.

Hughes Nursery. 5205 Nursery Rd. S.W., Cedar Rapids, IA 52401. (319) 396-7038. Balled-and-burlapped conifers and hardwoods; container conifers.

Inter-State. P.O. Box 208, Hamburg, IA 51640. 800-325-4180. Seeds.

Jordan's Nursery. P.O. Box 484, Hwy. 20 W., Cedar Falls, IA 50613. (319) 266-4717. All types of stock.

Landers' Nursery. R.R. 1, Box 19, Greene, IA 50636. (515) 823-5991. All types of conifers.

Leslie Gardens. P.O. Box 102, 502 S. Main St., Osceola, IA 50213. (515) 342-4408. Balled-and-burlapped, container, and tree-spade conifers and hardwoods.

Linn County Nursery. Hwy. 150, Center Point, IA 52213. (319) 849-1423. Bare-root hardwoods; balled-and-burlapped, container, and tree-spade conifers and hardwoods.

Linn Nursery. R.R. 1, Correctionville, IA 51016. (712) 372-4274. Bare-root, balled-and-burlapped, and tree-spade conifers and hardwoods.

Mallory Tree Transplanting. R.R. 2, Box 189, Milo, IA 50166. (515) 942-6380. Tree-spade conifers and hardwoods.

Maquoketa Nursery Sales. R.R. 1, Box 110, Maquoketa, IA 52060. (319) 652-3193. All types except tree-spade stock.

Merry Christmas Trees. 4200 First Ave. N.E., Cedar Rapids, IA 52402. (319) 393-8000. Tree-spade conifers.

Miller Nursery. R.R. 2, Box 61, Mediapolis, IA 52637. (319) 394-9296 (eve.). Balled-and-burlapped and container conifers and hardwoods.

Miller Nursery Co. 5155 N.W. 57th Ave., Des Moines, IA 50323. (515) 276-7505. All types of stock except bare-root conifers.

Mount Arbor Nurseries. 400 N. Center, Shenandoah, IA 51601. (712) 246-4250. Balled-and-burlapped conifers; container conifers and hardwoods; bare-root hardwoods.

Murphy's Walnut Hill Nursery. 1925 S.E. 82nd St., Runnells, IA 50237. (515) 262-6037. All types except container stock.

Nature's Way. R.R. 1, Box 62, Woodburn, IA 50275. (515) 342-6246. Woodland forbs.

Neldeberg Tree Spade Service. R.R. 1, Box 5, Whiting, IA 51063. (712) 458-2279. All types of stock.

N.I.A.C.C. (Horticulture). 500 College Dr., Mason City, IA 50401. (515) 421-4250. Bare-root hardwoods (ornamental species).

Oakwood Nursery. P.O. Box 44, Fairfield, IA 52556. (515) 472-6775. All types of stock.

Obrecht Tree Farm. P.O. Box 430, Malvern, IA 51551. (712) 624-8447. Balled-and-burlapped conifers.

Oelwein Landscaping. R.R. 1, Box 99, Oelwein, IA 50662. (319) 283-5578. All types of stock.

Pappas Landscaping Service. 1560 Plymouth Rd., Mason City, IA 50401. (515) 423-2703. All types except tree-spade stock.

Park Wholesale Nursery. 1107 Second Ave. N.W., Pocahontas, IA 50574. (712) 335-3259. Bare-root, balled-and-burlapped, and container conifers; balled-and-burlapped hardwoods.

P.B.S. The Landscapers. 5611 N.W. Beaver Dr., Des Moines, IA 50323. (515) 278-1118. Balled-and-burlapped, container, and tree-spade conifers and hardwoods.

Peckosh Nursery Services. R.R. 4, Box 855, Cedar Rapids, IA 52401. (319) 393-4486. Balled-and-burlapped and tree-spade conifers and hardwoods.

Peck's Green Thumb Nursery. 3990 Blairs Ferry Rd. N.E., Cedar Rapids, IA 52402. (319) 393-5946. All types of stock.

Pella Nursery Co. R.R. 3, Box 221, Pella, IA 50219. (515) 628-1285. Bare-root conifers and hardwoods in quantity; balled-and-burlapped and tree-spade conifers and hardwoods; container conifers.

Pellersels Tree Farm. 109 S. Seventh St., Sac City, IA 50583. (712) 662-7794. Balled-and-burlapped, container, and tree-spade conifers and hardwoods.

Pellett's Plant America. R.R. 2 on F48, Newton, IA 50208. (515) 792-7514. Balled-and-burlapped and container conifers and hardwoods.

Pilot Knob Nursery. R.R. 3, Clear Lake, IA 50428. (515) 357-3395. Bare-root, balled-and-burlapped, and container conifers.

Pine Acres. R.R. 3, New Hampton, IA 50659. (515) 394-4534. Bare-root seedlings in quantity; balled-and-burlapped conifers and hardwoods.

Platt's Nursery. 3700 University, Waterloo, IA 50701. (319) 234-2686. All types of stock.

Pork N' Pines. Gladbrook, IA 50635. (515) 473-2481. Balled-and-burlapped conifers.

Raccoon Valley Trees. 3420 S.W. 43rd St., Des Moines, IA 50321. (515) 285-4741. Tree-spade conifers.

Rees Landscaping. 215 Ashworth Dr., Waukee, IA 50263. (515) 987-4072. Balled-and-burlapped conifers and hardwoods.

Ridge Road Nursery. R.R. 2, Bellevue, IA 52031. (319) 588-4865. Bare-root, balled-and-burlapped, and container conifers and hardwoods.

Rodgers Nursery and Garden Center. 5135 Easton Blvd., Des Moines, IA 50317. (515) 262-5147. Bare-root hardwoods; balled-and-burlapped and container conifers and hardwoods.

Rosehill Nurseries. R.R. 2, Box 143, Panora, IA 50216. (515) 755-2152. All types of stock.

Rosemont Gardens. 4900 Nursery Rd. S.W., Cedar Rapids, IA 52401. (319) 396-7063. Balled-and-burlapped and container conifers.

Scotch Grove Nursery. Scotch Grove, IA 52331. (319) 465-3985. All types except tree-spade stock.

Shade Trees Today. R.R. 5, Boone, IA 50036. (515) 432-6465. Container and tree-spade conifers and hardwoods.

Sherman Nurseries. Charles City, IA 50616. Wholesale trees and seeds.

Smith Nursery Co. P.O. Box 515, Charles City, IA 50616. (515) 228-3239. Bare-root hardwoods and conifers; tree and shrub seeds.

State Forest Nursery. 2402 S. Duff Ave., Ames, IA 50010. (515) 233-1161. R.R. 1, Box 307A, Montrose, IA 52639. (319) 463-7167. Bare-root conifers and hardwoods in quantity; can only be used at a rural address, must not be resold as living trees, and cannot be used as windbreaks.

Strawberry Acres. R.R. 1, Marion, IA 52302. (319) 393-2366. Bare-root hardwoods in quantity; container hardwoods.

T & S Nursery. R.R. 3, Hawarden, IA 51023. (712) 552-1917. All types of stock.

Terry's Evergreen Valley Nursery. R.R. 2, Glenwood, IA 51534. (712) 527-4915. Bare-root, balled-and-burlapped, and tree-spade conifers; balled-and-burlapped and tree-spade hardwoods.

Tree Farm. R.R. 1, Box 230A, LeClaire, IA 52753. (319) 289-3060. Balled-and-burlapped and tree-spade conifers and hardwoods.

Tucker's Nursery. R.R. 1, New Hampton, IA 50659. (515) 394-2555. All types of conifers and tree-spade hardwoods.

Urbandale Nursery. 6900 Townsend, Urbandale, IA 50322. (515) 276-3131. All types of stock.

Walker Nursery. Walker, IA 52352. (319) 245-1091. Balled-and-burlapped, container, and tree-spade conifers.

Wark Tree Farm. R.R. 1, Indianola, IA 50125. (515) 961-3610. Bare-root and balled-and-burlapped conifers; tree-spade conifers and hardwoods.

Bob White Nursery. R.R. 10, West Des Moines, IA 50265. (515) 276-6429. Tree-spade conifers and hardwoods.

Williams Tree Farm. R.R. 1, Larchwood, IA 51241. (712) 477-2468. Bare-root, balled-and-burlapped, and container conifers and hardwoods.

Willow Creek Lawn and Garden Center. 6240 S.W. 63rd St., Des Moines, IA 50321. (515) 287-1537. Balled-and-burlapped and container conifers and hardwoods.

Zaiser's Nursery. 2400 Sunnyside Ave., Burlington, IA 52601. (319) 752-6871. Balled-and-burlapped, container, and tree-spade conifers and hardwoods.

Appendix 3.

WETLAND SEED AND PLANT SOURCES

County Road Greenhouses. R.R. 1, Box 74, Malta, IL 60150.

County Wetlands Nursery. P.O. Box 126, Muskego, WI 53150.

Kester's Wild Game Food Nurseries. 4488 Highway 116 East, P.O. Box V, Omro, WI 54963.

LaFayette Home Nursery. P.O. Box 1A, LaFayette, IL 61449.

Nature's Way. R.R. 1, Box 62, Woodburn, IA 50275. (515) 342-6246. Wetland forbs.

Prairie Bluestem Nursery. R.R. 2, Box 92, Hillboro, IL 62049.

Prairie Ridge Nursery. 9738 Overland Rd., R.R. 2, Mt. Horeb, WI 53572.

Wildlife Nurseries. P.O. Box 2724, Oshkosh, WI 54903.

Appendix 4.

DISTRICT FORESTERS

District 1. Allamakee, Clayton, and Dubuque counties. Janet Ott, P.O. Box 662, Elkader, IA 52043. (319) 245-1891.

District 2. Bremer, Butler, Cerro Gordo, Chickasaw, Fayette, Floyd, Howard, Mitchell, Winneshiek, and Worth counties. Gary Beyer, P.O. Box 4, Charles City, IA 50616. (515) 228-6611.

District 3. Benton, Black Hawk, Buchanan, Grundy, Iowa, Jasper, Marshall, Poweshiek, and Tama counties. Robert Hibbs, P.O. Box 681, Marshalltown, IA 50158. (515) 752-3352.

District 4. Cedar, Clinton, Delaware, Jackson, Johnson, Jones, and Linn counties. Steve Swinconos, P.O. Box 46, Anamosa, IA 52205. (319) 462-2768.

District 5. Des Moines, Lee, Louisa, Muscatine, and Scott counties. Stan Tate, 515 Townsend, Wapello, IA 52653. (319) 523-8319.

District 6. Davis, Henry, Jefferson, Keokuk, Van Buren, Wapello, and Washington counties. Ray Lehn, P.O. Box 568, Fairfield, IA 52556. (515) 472-2370.

District 7. Appanoose, Lucas, Mahaska, Marion, Monroe, and Wayne counties. Duane Bedford, R.R. 2, Box 310, Chariton, IA 50049. (515) 774-4918.

District 8. Boone, Carroll, Dallas, Greene, Guthrie, Madison, Polk, Story, and Warren counties. George Warford, 1918 Greene, Adel, IA 50003. (515) 993-4133.

District 9. Audubon, Cass, Fremont, Harrison, Mills, Montgomery, Page, Pottawattamie, and Shelby counties. Randy Goerndt, P.O. Box 152, Red Oak, IA 51566. (712) 623-4252.

District 10. Buena Vista, Cherokee, Crawford, Ida, Lyon, Monona, O'Brien, Osceola, Plymouth, Sac, Sioux, and Woodbury counties. Joe Schwartz, P.O. Box 65, LeMars, IA 51031. (712) 546-5161.

District 11. Adair, Adams, Clarke, Decatur, Ringgold, Taylor, and Union counties. Randy Goerndt, 500 E. Taylor, Creston, IA 50801. (515) 782-6761.

District 12. Calhoun, Clay, Dickinson, Emmet, Franklin, Hamilton, Hancock, Hardin, Humboldt, Kossuth, Palo Alto, Pocahontas, Webster, Winnebago, and Wright counties. Gail Kantak, 102 Eighth St. S., Humboldt, IA 50548. (515) 332-2761.

Bibliography

1. IOWA'S NATURAL LANDSCAPE COMMUNITIES

Bishop, R. A. 1981. Iowa's wetlands. *Proceedings of the Iowa Academy of Science* 88:11–16.

Bishop, R. A., and A. G. van der Valk. 1982. Wetlands. In Cooper.

Cooper, T. C., ed. 1982. *Iowa's Natural Heritage.* Des Moines: Iowa Academy of Science and Iowa Natural Heritage Foundation.

Farrar, D. R. 1981. Perspectives on Iowa's declining flora and fauna: A symposium. *Proceedings of the Iowa Academy of Science* 88:1.

Fenton, T. E., and G. Miller. 1982. Soils. In Cooper.

Graham, B. F., and D. Glenn-Lewin. 1982. Forests. In Cooper.

Great Plains Flora Association (T. M. Barkley, ed.). 1986. *Flora of the Great Plains.* Lawrence: University Press of Kansas.

Hayden, A. 1945. The selection of prairie areas in Iowa which should be preserved. *Proceedings of the Iowa Academy of Science* 52:127–148.

Kingsley, N. P. 1990. Personal communication. St. Paul: USDA Forest Service, North Central Forest Experiment Station.

Pammel, L. H. 1896. Iowa. *Proceedings of the American Forestry Association* 11:77–78.

Pearson, J. 1990. Personal communication. Des Moines: Preserves and Ecological Services Bureau, Iowa Department of Natural Resources.

Prior, J. C. 1991. *Landforms of Iowa.* Iowa City: University of Iowa Press.

Prior, J. C., R. G. Baker, G. R. Hallberg, and H. A. Semken. 1982. Glaciation. In Cooper.

Shimek, B. 1911. The prairies. *Bulletin of the Laboratory of Natural History* 6:169–240.

Smith, D. D. 1981. Iowa prairie: An endangered ecosystem. *Proceedings of the Iowa Academy of Science* 88:7–10.

Waite, P., and R. Shaw. 1982. Weather and climate. In Cooper.

2. PRAIRIE RESTORATION IN IOWA

Ahrenhoerster, R., and T. Wilson. 1981. *Prairie Restoration for the Beginner.* Des Moines: Prairie Seed Source.

————. n.d. Planting patterns. Typescript.

Aikman, J. M. 1949. What an academy can do to promote the conservation of natural resources. *Proceedings of the Iowa Academy of Science* 56:29–36.

Anderson, W. A. 1936. Progress in the regeneration of the prairie at Lakeside Laboratory. *Proceedings of the Iowa Academy of Science* 43:87–93.

————. 1945. On transplanting prairie species. *Proceedings of the Iowa Academy of Science* 52:93–94.

Baker, R. G., and K. L. van Zant. 1978. The history of prairie in northwest Iowa: The pollen and plant macrofossil record. In Glenn-Lewin and Landers.

Becic, J. N., and T. B. Bragg. 1978. Grassland reestablishment in eastern Nebraska using burning and mowing management. In Glenn-Lewin and Landers.

Betz, R. F. 1986. One decade of research in prairie restoration at Fermi National Accelerator Laboratory, Batavia, Illinois. In Clambey and Pemble.

Braband, L. 1986. Railroad grasslands as bird and mammal habitats in central Iowa. In Clambey and Pemble.

Bragg, T. B. 1978. Allwine Prairie Preserve: A reestablished bluestem grassland research area. In Glenn-Lewin and Landers.

Christiansen, P. A. 1990. Personal communication. Mount Vernon, Iowa: Department of Biology, Cornell College.

Christiansen, P. A., and R. Q. Landers. 1966. Notes on prairie species in Iowa I. *Proceedings of the Iowa Academy of Science* 73:51–59.

————. 1969. Notes on prairie species in Iowa II. *Proceedings of the Iowa Academy of Science* 76:94–104.

Clambey, G. K., and R. H. Pemble, eds. 1986. *Proceedings of the Ninth North American Prairie Conference.* Fargo, N.D.: Tri-College University.

Coffin, L. S. 1902. Breaking the prairie. *Annals of Iowa* 5:447–452.

Costello, D. F. 1969. *The Prairie World.* New York: T. Y. Crowell.

Cox, C. A. 1987. Evaluation of three prairie restorations. *Restoration and Management Notes* 5 (1): 25–26.

Crane, J., and G. W. Olcott. 1933. *Report on the Iowa Twenty-five Year Conservation Plan.* Des Moines: Iowa Board of Conservation and Iowa Fish and Game Commission.

Crist, A., and D. C. Glenn-Lewin. 1978. The structure of community and environmental gradients in a northern Iowa prairie. In Glenn-Lewin and Landers.

Curtis, J. T. 1959. *Vegetation of Wisconsin.* Madison: University of Wisconsin Press.

Dancer, W. S. 1985. Prairie soil restoration research summarized (Illinois). *Restoration and Management Notes* 3 (1): 30–31.

Diboll, N. 1986. Mowing as an alternative to burning for control of cool season exotic grasses in prairie grass plantings. In Clambey and Pemble.

Drake, L. D. 1978. Prairie models for agricultural systems. In Glenn-Lewin and Landers.

Duvick, D. N., and T. J. Blasing. 1981. A dendroclimatic reconstruction of annual precipitation amounts in Iowa since 1680. *Water Resources Research* 17: 1183–1189.

Eddelman, L. E., and P. L. Meinhardt. 1981. Seed viability and seedling vigor in selected prairie plants. In Stuckey and Reese.

Edwards, T. 1978. Buffalo and prairie ecology. In Glenn-Lewin and Landers.

Eilers, L. J. n.d. Vascular plants of Iowa. Typescript.

Farrar, D. R. 1989. Personal communication. Ames: Department of Botany, Iowa State University.

Fenton, T. E. 1983. Mollisols. In *Pedogenesis and Soil Taxonomy*, vol. 2: *The Soil Orders*, edited by L. P. Wilding, N. E. Smeck, and G. F. Hall. New York: Elsevier Science Publishing Co.

Freckmann, R. W. 1966. The prairie remnants of the Ames area. *Proceedings of the Iowa Academy of Science* 73: 126–136.

George, R. R. 1978. Native grass pastures as nesting habitat for bobwhite quail and ring-necked pheasant. In Glenn-Lewin and Landers.

Glenn-Lewin, D. C., and R. Q. Landers, eds. 1978. *Fifth Midwest Prairie Conference Proceedings*. Ames: Iowa State University.

Glenn-Lewin, D. C., R. H. Laushman, and P. D. Whitson. 1984. The vegetation of the Paleozoic Plateau, northeastern Iowa. *Proceedings of the Iowa Academy of Science* 91: 22–27.

Graham, B. F. 1975. CERA: An outdoor biological laboratory. In Wali.

Hayden, A. 1945. The selection of prairie areas in Iowa which should be preserved. *Proceedings of the Iowa Academy of Science* 52: 127–148.

———. 1946. A progress report on the preservation of prairie. *Proceedings of the Iowa Academy of Science* 53: 45–82.

Hill, G. R., and W. J. Platt. 1975. Some effects of fire upon a tallgrass prairie plant community in northwestern Iowa. In Wali.

Howe, R. H. 1984. Wings over the prairie. *Iowa Conservationist* 43 (9): 5–7.

Hulbert, L. C. 1978. Controlling experimental bluestem fires. In Glenn-Lewin and Landers.

———. 1986. Fire effects on tallgrass prairie. In Clambey and Pemble.

Iowa Department of Natural Resources. 1988. *Prairie Establishment and*

Management Guide. Des Moines: Office of County Conservation Activities, Iowa Department of Natural Resources.

Jastrow, J. D. 1987. Changes in soil aggregation associated with tallgrass prairie restoration. *American Journal of Botany* 74 (11): 1656–1664.

Kaduce, R. 1990. Personal communication. Ledges State Park: Iowa Department of Natural Resources.

Landers, R. Q., and P. Christiansen. n.d. Notes on starting a prairie: Our experience in Iowa. Typescript.

Lebovitz, R. 1987. The prairie: A model and metaphor for sustainable agriculture. *Land Report* 31:18–19.

Lekwa, S. 1984. Prairie restoration and management. *Iowa Conservationist* 43 (9): 12–14.

————. 1990. Personal communication. Colo, Iowa: Story County Conservation Board.

Levenson, J. B. 1981. Woodlots as biogeographic islands in southeastern Wisconsin. In *Forest Island Dynamics on Man-Dominated Landscapes*, edited by R. L. Burgess and D. M. Sharpe. New York: Springer-Verlag.

Liegel, K., and J. Lyon. 1986. Prairie restoration program at the International Crane Foundation. In Clambey and Pemble.

McClain, W. E. 1986. *Illinois Prairie: Past and Future. A Restoration Guide*. Springfield: Division of Natural Heritage, Illinois Department of Conservation.

Mutel, C. F. 1989. *Fragile Giants: A Natural History of the Loess Hills*. Iowa City: University of Iowa Press.

Nelson, H. L. 1986. Prairie restoration in the Chicago area. *Restoration and Management Notes* 5 (2): 60–67.

Nichols, S., and L. Entine. 1976. *Prairie Primer*. Madison: University of Wisconsin-Extension.

Nuzzo, V. 1978. Propagation and planting of prairie forbs and grasses in southern Wisconsin. In Glenn-Lewin and Landers.

————. 1981. Criteria for introduction of species to natural areas. In Stuckey and Reese.

Pauly, W. R. 1982. *How to Manage Small Prairie Fires*. Madison: Dane County Highway and Transportation Department.

Pearson, J. 1990. Personal communication. Des Moines: Preserves and Ecological Services Bureau, Iowa Department of Natural Resources.

Piper, J. 1988. New roots for agriculture: 1988 research season underway. *Land Report* 32:16–17.

Platt, W. J. 1975. Vertebrate fauna of the Cayler Prairie Preserve, Dickinson

County, Iowa. *Proceedings of the Iowa Academy of Science* 82:106–108.

Reed, D. M., and J. A. Schwarzmeier. 1978. The prairie corridor concept: Possibilities for planning large-scale preservation and restoration. In Glenn-Lewin and Landers.

Richards, M. S., and R. Q. Landers. 1973. Responses of species in Kalsow Prairie, Iowa, to an April fire. *Proceedings of the Iowa Academy of Science* 80:159–161.

Ripp, M. 1985. Native vegetation reintroduction efforts using 2 and 3 year old rootstock. *Restoration and Management Notes* 3 (1): 33.

Risser, P. G., E. C. Birney, H. D. Blocker, S. W. May, W. J. Parton, and J. A. Wiens. 1981. *The True Prairie Ecosystem.* U.S./I.B.P. Synthesis Series 16. Stroudsburg, Pa.: Hutchinson Ross Publishing Co.

Rock, H. W. 1981. *Prairie Propagation Handbook.* 6th ed. Milwaukee: Wehr Nature Center, Whitnall Park.

Roosa, D. M. 1978. Prairie preservation in Iowa: History, present status, and future plans. In Glenn-Lewin and Landers.

Sauer, C. O. 1950. Grassland, climates, fires, and man. *Journal of Range Management* 3:16–22.

Schennum, W. E. 1986. A comprehensive survey for prairie remnants in Iowa: Methods and preliminary results. In Clambey and Pemble.

Schramm, P. 1978. The "do's and dont's" of prairie restoration. In Glenn-Lewin and Landers.

Schramm, P., and R. L. Kalvin. 1978. The use of prairie in strip mine reclamation. In Glenn-Lewin and Landers.

Shimek, B. 1911a. The prairies. *Bulletin of the Laboratory of Natural History of Iowa* 6 (2): 169–240.

———. 1911b. The prairies. *Bulletin of the Laboratory of Natural History of Iowa* 7 (2): 3–69.

———. 1925. The persistence of the prairie. *University of Iowa Studies in Natural History* 11 (5): 3–24.

———. 1931. Relation between migrant and native flora of the prairie region. *University of Iowa Studies in Natural History* 14 (2): 10–16.

Skinner, R. M. 1975. Grassland use patterns and prairie bird populations in Missouri. In Wali.

Smith, D. D. 1981. Iowa prairie: An endangered ecosystem. *Proceedings of the Iowa Academy of Science* 88:7–10.

Smith, D. D., and P. Christiansen. 1982. Prairies. In *Iowa's Natural Heritage,* edited by T. C. Cooper. Des Moines: Iowa Academy of Science and Iowa Natural Heritage Foundation.

Soil and Water Conservation Society. 1987. *Sources of Native Seeds and Plants.* Ankeny, Iowa.

Stuckey, R. L., and K. J. Reese, eds. 1981. *Proceedings of the Sixth North American Prairie Conference, Ohio Biological Survey Notes* 15. Columbus: Ohio Biological Survey.

Transeau, E. N. 1905. Climatic centers and centers of plant distribution. *Annual Report of the Michigan Academy of Science* 7:73–75.

———. 1935. The prairie peninsula. *Ecology* 16:423–437.

Twarok, C. J. 1989. Personal communication. Ames: Department of Forestry, Iowa State University.

USDA. 1975. *Soil Taxonomy.* Agriculture Handbook 436. Washington, D.C.: Government Printing Office.

Wagner, R. H. 1975. The American Prairie Inventory: A preliminary report. In Wali.

Wali, M. K., ed. 1975. *Proceedings of the Fourth Midwest Prairie Conference.* Grand Forks: University of North Dakota Press.

Weaver, J. E. 1954. *North American Prairie.* Lincoln, Nebr.: Johnson Publishing Co.

White, J. A., and D. C. Glenn-Lewin. 1984. Regional and local variation in tallgrass prairie remnants of Iowa and eastern Nebraska. *Vegetatio* 57:65–78.

Widstrand, S. 1985. Costs compared for burning vs. mowing on small prairies and old fields. *Restoration and Management Notes* 3 (1): 31.

Woehler, E. E., and M. A. Martin. 1978. Establishment of prairie grasses and forbs with the use of herbicides. In Glenn-Lewin and Landers.

Wooley, J. 1984. Prairie chicken update. *Iowa Conservationist* 43 (9): 10–11.

3. FOREST RESTORATION IN IOWA

Aikman, J. M., and A. W. Smelser. 1938. The structure and environment of forest communities in central Iowa. *Ecology* 19:141–150.

Anderson, B. 1990. Personal communication. Springfield: Illinois Nature Preserves Commission.

Anderson, R. C., and L. E. Brown. 1986. Stability and instability in plant communities following fire. *American Journal of Botany* 73 (3): 364–368.

Ashby, W. C. 1987. Forests. In *Restoration Ecology: A Synthetic Approach to Ecological Research,* edited by W. R. Jordan III, M. E. Gilpin, and J. D. Aber. Cambridge: Cambridge University Press.

Barrett, J. W. 1980. *Regional Silviculture of the United States.* New York: J. Wiley and Sons.

Beattie, M., C. Thompson, and L. Levine. 1983. *Working with Your Woodland: A Landowners Guide.* Hanover, N.H.: University Press of New England.

Bowles, J. B. 1981. Iowa's mammal fauna: An era of decline. *Proceedings of the Iowa Academy of Science* 88:38–42.

Brinson, M. M., B. L. Swift, R. C. Plantico, and J. S. Barclay. 1981. *Riparian Ecosystems: Their Ecology and Status.* Biological Service Program FWS/OBS-81/17. Washington, D.C.: U.S. Fish and Wildlife Service.

Bromley, P., J. Starr, J. Sims, and D. Coffman. 1990. *A Landowner's Guide to Wildlife Abundance through Forestry.* Virginia Cooperative Extension Service Pub. 420-138. Blacksburg: Virginia Polytechnic Institute and State University.

Brown, S., M. Brinson, and A. Lugo. 1978. Structure and function of riparian wetlands. In *Strategies for Protection and Management of Floodplain Wetlands and Other Riparian Systems,* edited by R. Johnson and J. McCormick. General Technical Report WO-12. Washington, D.C.: USDA Forest Service.

Burel, F., and J. Baudry. 1989. Hedgerow network patterns and processes in France. In Zonneveld and Forman.

Burgess, R. L., and D. M. Sharpe, eds. 1981. *Forest Island Dynamics in Man-Dominated Landscapes.* New York: Springer-Verlag.

Capel, S. W. 1988. Design of windbreaks for wildlife in the great plains. *Agriculture, Ecosystems, and Environment* 22/23:337–347.

Cooper, T. C., ed. 1982. *Iowa's Natural Heritage.* Des Moines: Iowa Academy of Science and Iowa Natural Heritage Foundation.

Crane, J., and G. W. Olcott. 1933. *Report on the Iowa Twenty-five Year Conservation Plan.* Des Moines: Iowa Board of Conservation and Iowa Fish and Game Commission.

Cubbage, F. W., and J. E. Gunter. 1987. Conservation Reserves: Can they promote tree crops while protecting erodible soil? *Journal of Forestry* 85 (4): 21–27.

Davidson, R. A. 1960. Plant communities of southeastern Iowa. *Proceedings of the Iowa Academy of Science* 67:162–173.

Diekelmann, J., and R. Schuster. 1982. *Natural Landscaping: Designing with Native Plant Communities.* New York: McGraw-Hill.

Dinsmore, J. J. 1981. Iowa's avifauna: Changes in the past and prospects for the future. *Proceedings of the Iowa Academy of Science* 88:28–37.

Dinsmore, J. J., T. H. Kent, D. Koenig, P. C. Petersen, and D. M. Roosa. 1984. *Iowa Birds.* Ames: Iowa State University Press.

Dorney, C. H., and J. R. Dorney. 1989. An unusual oak savanna in northeastern Wisconsin: The effect of Indian-caused fire. *American Midland Naturalist* 122:103–113.

Dorney, R. S. 1983. Costs of woodland, lawn restoration and maintenance compared. *Restoration and Management Notes* 1 (1): 22–23.

Eilers, L. J. 1982. Iowa as it was. In Cooper.

———. n.d. Vascular plants of Iowa. Typescript.

Fazio, J. R. 1985. *The Woodland Steward.* Moscow, Idaho: Woodland Press.

Forbes, R. D. 1971. *Woodlands for Profit and Pleasure.* Washington, D.C.: American Forestry Association.

Forman, R. T. T. 1983. Corridors in a landscape: Their ecological structure and function. *Ekologia* (C.S.S.R.) 2:375–387.

Gottfried, B. M. 1979. Small mammal populations in woodlot islands. *American Midland Naturalist* 102:105–109.

Graham, B. F., and D. Glenn-Lewin. 1982. Forests. In Cooper.

Harms, W. B., and P. Opdam. 1989. Woods as habitat patches for birds: Application in landscape planning in the Netherlands. In Zonneveld and Forman.

Haynes, R. J., and L. Moore. 1988. Reestablishment of bottomland hardwoods within national wildlife refuges in the southeast. In *Increasing Our Wetland Resources,* edited by J. Zelazny and J. S. Feierabend. Washington, D.C.: National Wildlife Federation.

Hightshoe, G. 1978. *Native Trees for Urban and Rural America.* Ames: Iowa State University Research Foundation.

———. 1984. Computer-assisted program for forest preservation/conservation/restoration: Upper midwest region. *Landscape Journal* 1: 45–60.

Howell, E. A. 1986. Woodland restoration: An overview. *Restoration and Management Notes* 4 (1): 13–17.

Iowa Department of Natural Resources. n.d. Weed control for tree and shrub seedlings. Des Moines.

Iowa State Planning Board. 1935. *Iowa Twenty-five Year Conservation Plan.* Des Moines: Iowa State Conservation Commission.

Johnson, R. J., and M. M. Beck. 1988. Influences of shelterbelts on wildlife management and biology. *Agriculture, Ecosystems, and Environment* 22/23:301–335.

Kingsley, N. 1990. Personal communication. St. Paul: USDA Forest Service, North Central Forest Experiment Station.

Lea, R. 1988. Management of eastern United States bottomland hardwood forests. In *The Ecology and Management of Wetlands*, vol. 2: *Management of Wetlands*, edited by D. D. Hook, W. H. McKee, Jr., H. K. Smith, J. Gregory, V. G. Burrell, Jr., M. R. DeVoe, R. E. Sojka, S. Gilbert, R. Banks, L. H. Stolzy, C. Brooks, T. D. Matthews, and T. D. Shear. Portland: Timber Press.

Levenson, J. B. 1981. Woodlots as biogeographic islands in southeastern Wisconsin. In Burgess and Sharpe.

Lovejoy, T. M., and D. C. Oren. 1981. The minimum critical size of ecosystems. In Burgess and Sharpe.

MacClintock, L., R. F. Whitcomb, and B. L. Whitcomb. 1977. Evidence for the value of corridors and minimization of isolation in preservation of biotic diversity. *American Bird* 31 (1): 6–16.

Matthiae, P. E., and F. Stearns. 1981. Mammals in forest islands in southeastern Wisconsin. In Burgess and Sharpe.

Merriam, G. 1989. Ecological processes in the time and space of farmland mosaics. In Zonneveld and Forman.

Merritt, C. 1980. The central region. In Barrett.

Middleton, J., and G. Merriam. 1983. Distribution of woodland species in farmland woods. *Journal of Applied Ecology* 20:625–639.

Minckler, L. S. 1975. *Woodland Ecology: Environmental Forestry for the Small Owner.* Syracuse: Syracuse University Press.

Missouri Department of Conservation. n.d. Managing the streamside forest. Jefferson City.

Niemann, D. A., and R. Q. Landers, Jr. 1974. Forest communities in Woodman Hollow State Preserve, Iowa. *Proceedings of the Iowa Academy of Science* 81:176–184.

Oliver, C. D., and T. M. Hinckley. 1987. Species, stand structures, and silvicultural manipulation patterns for the streamside zone. In *Streamside Management: Forestry and Fishery Interactions*, edited by E. O. Salo and T. W. Cundy. Seattle: University of Washington.

Opdam, P., G. Rijsdijk, and F. Hustings. 1985. Bird communities in small woods in an agricultural landscape: Effects of area and isolation. *Biological Conservation* 34:333–352.

Packard, S. 1988. Just a few oddball species: Restoration and the rediscovery of tallgrass savanna. *Restoration and Management Notes* 6 (1): 13–20.

Pammel, L. H. 1896. Iowa. *Proceedings of the American Foresters Association* 11:77–78.

Peterjohn, W. T., and D. L. Correll. 1984. Nutrient dynamics in an agricultural watershed: Observation on the role of a riparian forest. *Ecology* 65:1466–1475.

Prior, J. C., R. G. Baker, G. R. Hallberg, and H. A. Semken. 1982. Glaciation. In Cooper.

Robbins, C. S. 1979. Effects of forest fragmentation on bird populations. In *Management of North Central and Northeastern Forests for Non-game Birds*. General Technical Report NC-51. Washington, D.C.: USDA Forest Service.

Roosa, D. M. 1982. Natural regions of Iowa. In Cooper.

Rust, R. H. 1983. Alfisols. In *Pedogenesis and Soil Taxonomy*, vol. 2: *The Soil Orders*, edited by L. P. Wilding, N. E. Smeck, and G. F. Hall. New York: Elsevier Science Publishing Co.

Shimek, B. 1899. The distribution of forest trees in Iowa. *Proceedings of the Iowa Academy of Science* 7:47–59.

Smith, D. W., and N. E. Linnartz. 1980. The southern hardwood region. In Barrett.

Society of American Foresters (F. C. Ford-Robinson and R. K. Winters, eds.). 1983. *Terminology of Forest Science Technology, Practice, and Products*. Washington, D.C.

Spurr, S. H., and B. V. Barnes. 1980. *Forest Ecology*. 3d ed. New York: J. Wiley and Sons.

Stauffer, D. F., and L. B. Best. 1980. Habitat selection by birds of riparian communities: Evaluating effects of habitat alterations. *Journal of Wildlife Management* 44:1–15.

Thomson, G. W. 1987. Iowa's forest area in 1832: A reevaluation. *Proceedings of the Iowa Academy of Science* 94:116–120.

Thomson, G. W., and H. G. Hertel. 1981. The forest resources of Iowa in 1980. *Proceedings of the Iowa Academy of Science* 88:2–6.

Tregay, R., and D. Moffatt. 1980. An ecological approach to landscape design and management in Oakwood, Warrington. *Landscape Design* 132:33–36.

Turner, R. E., S. W. Forsythe, and N. J. Craig. 1981. Bottomland hardwood forest land resources of the southeastern United States. In *Wetlands of Bottomland Hardwood Forests*, edited by J. R. Clark and J. Benforado. New York: Elsevier Science Publishing Co.

Twarok, C. J. 1989. Personal communication. Ames: Department of Forestry, Iowa State University.

USDA. 1974. *Seeds of Woody Plants in the United States*. Agriculture Handbook 450. Washington, D.C.: Government Printing Office.

———. 1975. *Soil Taxonomy*. Agricultural Handbook 436. Washington, D.C.: Government Printing Office.

USDA Forest Service. 1980. *Forest Inventory of Iowa*. Washington, D.C.: Government Printing Office.

van der Linden, P. J., and D. R. Farrar. 1984. *Forest and Shade Trees of Iowa*. Ames: Iowa State University Press.

Wade, G. L. 1989. Grass competition and establishment of native species from forest soil seed banks. *Landscape and Urban Planning* 17:135–149.

White, A. S. 1983. The effects of thirteen years of annual prescribed burning on a *Quercus ellipsoidalis* community in Minnesota. *Ecology* 64:1081–1085.

Winebar, B., and J. Gunter. 1984. *Costs of Forest Management Practices in the Lake States*. Michigan State University Agricultural Experiment Station Research Report 457. East Lansing: Michigan State University.

Wray, P. H. 1986. *Growing Seedlings from Seed*. Forestry Extension Notes F-304. Ames: Iowa State University Cooperative Extension Service.

———. 1987. *Management of Floodplain Forests*. Forestry Extension Notes F-326. Ames: Iowa State University Cooperative Extension Service.

———. 1988. *Tree Planting in Iowa*. Bulletin Pm-496. Ames: Iowa State University Cooperative Extension Service.

———. 1989. *Woodland Management: Improving Woodlands*. Bulletin Pm-1374c. Ames: Iowa State University Cooperative Extension Service.

———. 1990. *Woodland Management: Harvesting and Regenerating Upland Hardwoods*. Bulletin Pm-1374d. Ames: Iowa State University Cooperative Extension Service.

Wray, P. H., and W. Farris. 1989. *Woodland Management: Understanding Trees and Woodlands*. Bulletin Pm-1374a. Ames: Iowa State University Cooperative Extension Service.

Wray, P. H., and R. C. Schultz. 1988. *Fall Planting of Bare-root Seedlings*. Forestry Extension Notes F-355. Ames: Iowa State University Cooperative Extension Service.

Zonneveld, I. S., and R. T. T. Forman, eds. 1989. *Changing Landscapes: An Ecological Perspective*. New York: Springer-Verlag.

4. WETLAND RESTORATION IN IOWA

Adamus, P. R., and L. T. Stockwell. 1983. *A Method for Wetland Functional Assessment*. Report F.H.W.A.P-82-83. Washington, D.C.: U.S. Department of Transportation, Federal Highway Administration.

Baskett, R. K. 1988. Grand Pass Wildlife Area, Missouri: Modern wetland restoration strategy at work. In Zelazny and Feierabend.

Bastian, R. K., and J. Benforado. 1988. Water quality functions of wetlands: Natural and managed systems. In Hook et al., vol. 1.

Bishop, R. A. 1981. Iowa's wetlands. *Proceedings of the Iowa Academy of Science* 88:11–16.

——. 1989. Personal communication. Des Moines: Iowa Conservation Commission.

Bishop, R. A., and A. G. van der Valk. 1982. Wetlands. In *Iowa's Natural Heritage,* edited by T. C. Cooper. Des Moines: Iowa Academy of Science and Iowa Natural Heritage Foundation.

Brinson, M. M., B. L. Swift, R. C. Plantico, and J. S. Barclay. 1981. *Riparian Ecosystems: Their Ecology and Status.* Biological Service Program FWS/OBS-81/17. Washington, D.C.: U.S. Fish and Wildlife Service.

Brown, M., and J. J. Dinsmore. 1986. Implications of marsh size and isolation for marsh bird management. *Journal of Wildlife Management* 50:392–397.

Buol, S. A., and R. A. Rebertus. 1988. Soil formation under hydromorphic conditions. In Hook et al., vol. 1.

Campbell, K. L., and H. P. Johnson. 1975. Hydraulic simulation of watersheds with artificial drainage. *Water Resources Research* 11:120–126.

Carter, V., and R. P. Novitsky. 1988. Some comments on the relation between groundwater and wetlands. In Hook et al., vol. 1.

Christiansen, J. L. 1981. Population trends among Iowa's amphibians and reptiles. *Proceedings of the Iowa Academy of Science* 88:24–27.

Cowardin, L. M., V. Carter, F. C. Golet, and E. T. LaRoe. 1979. *Classification of Wetlands and Deepwater Habitats of the United States.* Biological Service Program FWS/OBS-79/31. Washington, D.C.: U.S. Fish and Wildlife Service.

Davis, C. B., J. L. Baker, A. G. van der Valk, and C. E. Beer. 1981. Prairie pothole marshes as traps for nitrogen and phosphorus in agricultural runoff. *Proceedings of the Midwest Conference on Wetland Values and Management,* edited by B. Richardson. Navarre, Minn.: Freshwater Society.

Dinsmore, J. J. 1981. Iowa's avifauna: Changes in the past and prospects for the future. *Proceedings of the Iowa Academy of Science* 88:28–37.

Dinsmore, J. J., T. H. Kent, D. Koenig, P. C. Petersen, and D. M. Roosa. 1984. *Iowa Birds.* Ames: Iowa State University Press.

Erlandson, C. S. 1987. The potential role of seed banks in the restoration of drained prairie wetlands. M.S. thesis. Ames: Iowa State University.

Fischel, M. 1988. Wetland restoration/creation and the controversy over its use in mitigation: An introduction. In Zelazny and Feierabend.

Fischer, W. 1990. Personal communication. Rock Island, Ill.: U.S. Fish and Wildlife Service.

Fritzell, E. K. 1988. Mammals and wetlands. In Hook et al., vol. 1.

Haynes, R. J., and L. Moore. 1988. Reestablishment of bottomland hard-woods within national wildlife refuges in the southeast. In Zelazny and Feierabend.

Heiser, N. 1989. Personal communication. Spirit Lake, Iowa: Iowa Department of Natural Resources.

Hey, D. L. 1988. Des Plaines River Wetlands Demonstration Project: Developing and Implementing Goals and Objectives. In Zelazny and Feierabend.

Hook, D. D. 1988. Criteria for creating and restoring forested wetlands in the southern United States. In Zelazny and Feierabend.

Hook, D. D., W. H. McKee, Jr., H. K. Smith, J. Gregory, V. G. Burrell, Jr., M. R. DeVoe, R. E. Sojka, S. Gilbert, R. Banks, L. H. Stolzy, C. Brooks, T. D. Mathews, and T. D. Shear, eds. 1988. *The Ecology and Management of Wetlands*, vol. 1: *Ecology of Wetlands*. Portland: Timber Press.

————. 1988. *The Ecology and Management of Wetlands*, vol. 2: *Management of Wetlands*. Portland: Timber Press.

Howell, E. A. 1986. Woodland restoration: An overview. *Restoration and Management Notes* 4 (1): 13–17.

Hubbard, D. E. 1988. *Glaciated Prairie Wetland Functions and Values: A Synthesis of the Literature*. Biological Report 88/43. Washington, D.C.: U.S. Fish and Wildlife Service.

Iowa Department of Natural Resources. 1988. Acquisition and restoration of wetlands in 31 Iowa counties. Typescript. Des Moines.

ISPAID. 1989. *Iowa Soil Properties and Interpretations Database, Version 4.1*. Ames: Iowa State University.

Kadlec, R. H. 1988. Monitoring wetland responses. In Zelazny and Feierabend.

Kantrud, H. A. 1986. *Effects of Vegetation Manipulation on Breeding Waterfowl in Prairie Wetlands: A Literature Review*. Fish and Wildlife Technical Report 3. Washington, D.C.: U.S. Fish and Wildlife Service.

Kantrud, H. A., and R. E. Stewart. 1977. Use of natural basin wetlands by breeding waterfowl in North Dakota. *Journal of Wildlife Management* 41:243–253.

Kantrud, H. A., J. B. Millar, and A. G. van der Valk. 1989. Vegetation of wetlands of the prairie pothole region. In *Northern Prairie Wetlands*, edited by A. G. van der Valk. Ames: Iowa State University Press.

Kittelson, J. M. 1988. Analysis of flood peak moderation by depressional wetland sites. In Hook et al., vol. 1.

LaGrange, T. G., and J. J. Dinsmore. 1989. Plant and animal community responses to restored Iowa wetlands. *Prairie Naturalist* 21 (1): 39–48.

Lammers, T. G., and A. G. van der Valk. 1977. A checklist of the aquatic and wetland vascular plants of Iowa I. *Proceedings of the Iowa Academy of Science* 84 : 41–88.

———. 1979. A checklist of the aquatic and wetland vascular plants of Iowa II. *Proceedings of the Iowa Academy of Science* 85 : 121–163.

Larson, J. S. 1988. Wetland creation and restoration: An outline of the scientific perspective. In Zelazny and Feierabend.

Lekwa, S. 1989. Personal communication. Colo, Iowa: Story County Conservation Board.

Leoschke, M., and J. Pearson. 1988. Fen: A special kind of wetland. *Iowa Conservationist* 47 : 16–19.

Leventhal, E. 1990. Alternative uses of wetlands other than conventional farming in Iowa, Kansas, Missouri, and Nebraska. Report to USEPA National Network for Environmental Management Studies, Region 7, Kansas City, Kansas.

Madsen, C. R. 1988. Wetland restoration in western Minnesota. In Zelazny and Feierabend.

Maltby, E. E. 1988. Wetland resources and future prospects: An international perspective. In Zelazny and Feierabend.

Mann, G. E. 1955. *Wetlands Inventory of Iowa.* Washington, D.C.: Office of River Basin Studies, U.S. Fish and Wildlife Service.

Menzel, B. W. 1981. Iowa's waters and fishes: A century and a half of change. *Proceedings of the Iowa Academy of Science* 88 : 17–23.

Merezheko, A. I. 1973. Role of higher aquatic plants in self-purification of lakes and streams. *Hydrobiology Journal* 9 : 103–109.

Murkin, H. R., and D. A. Wrubleski. 1988. Aquatic invertebrates of freshwater wetlands: Function and ecology. In Hook et al., vol. 1.

Neiring, W. A. 1988. Endangered, threatened, and rare wetland plants and animals of the continental U.S. In Hook et al., vol. 1.

Odum, W. E. 1988. Predicting ecosystem development following creation and restoration of wetlands. In Zelazny and Feierabend.

Parker, W. B. 1988. Use of hydric soils to assist in worldwide wetland inventories. In Hook et al., vol. 2.

Pederson, R. L., and L. M. Smith. 1988. Implications of wetland seedbank research: A review of great basin and prairie marsh studies. In *Interdisciplinary Approaches to Freshwater Wetlands Research*, edited by D. A. Wilcox. East Lansing: Michigan State University Press.

Roosa, D. M., and L. J. Eilers. 1978. *Endangered and Threatened Iowa Vascular Plants*. Special Report 5. Des Moines: State Preserves Advisory Board.

Sather, J. M., and R. D. Smith. 1984. *An Overview of Major Wetland Functions and Values*. FWS/OBS-84/18. Washington, D.C.: U.S. Fish and Wildlife Service.

Shaw, S. P., and C. G. Fredine. 1956. *Wetlands of the United States*. Office of River Basin Studies Circular 39. Washington, D.C.: U.S. Fish and Wildlife Service.

Siegel, D. I. 1988. A review of the recharge-discharge function of wetlands. In Hook et al., vol. 1.

Smith, D. W., and N. E. Linnartz. 1980. The southern hardwood region. In *Regional Silviculture of the United States*, edited by J. W. Barrett. New York: J. Wiley and Sons.

Tiner, R. W., Jr. 1984. *Wetlands of the United States: Current Status and Recent Trends*. Washington, D.C.: National Wetlands Inventory, U.S. Fish and Wildlife Service.

Turner, R. E., S. W. Forsythe, and N. J. Craig. 1981. Bottomland hardwood forest land resources of the southeastern United States. In *Wetlands of Bottomland Hardwood Forests*, edited by J. R. Clark and J. Benforado. New York: Elsevier Science Publishing Co.

Twarok, C. J. 1989. Personal communication. Ames: Department of Forestry, Iowa State University.

USDA 1984. *Soil Conservation Service Specifications for Wildlife Wetland Habitat in Iowa*, Specification 644. Washington, D.C.

USDA (National Technical Committee for Hydric Soils). 1987. *Hydric Soils of the United States*. Washington, D.C.: USDA Soil Conservation Service.

van der Valk, A. G. 1981. Succession in wetlands: A Gleasonian approach. *Ecology* 62:688–696.

———. 1985. Vegetation dynamics of prairie glacial marshes. In *The Population Structure of Vegetation*, edited by J. White. Dordrecht: W. Junk.

van der Valk, A. G., and C. B. Davis. 1978. The role of seed banks in vegetation dynamics of prairie glacial marshes. *Ecology* 59:322–335.

Weinhold, C. E., and A. G. van der Valk. 1989. The impact of duration of drainage on the seed banks of northern prairie wetlands. *Canadian Journal of Botany* 67:1878–1884.

Weller, M. W., and C. E. Spatcher. 1965. *Role of Habitat in the Distribution and Abundance of Marsh Birds.* Special Report 43. Ames: Home Economics Experiment Station, Iowa State University.

Wray, P. H. 1988. *Tree Planting in Iowa.* Bulletin Pm-496. Ames: Iowa State University Cooperative Extension Service.

——. 1989. Personal communication. Ames: Department of Forestry, Iowa State University.

Wray, P. H., and R. C. Schultz. 1988. *Fall Planting of Bare-root Seedlings.* Forestry Extension Notes F-355. Ames: Iowa State University Cooperative Extension Service.

Zedler, J. B. 1988. Why it's so difficult to replace lost wetland functions. In Zelazny and Feierabend.

Zelazny, J., and J. S. Feierabend, eds. 1988. *Increasing Our Wetland Resources.* Washington, D.C.: National Wildlife Federation.

APPENDIXES

Drake, L. D. 1990. Tallgrass prairie sources. Typescript.

Hildebrandt, R., P. H. Wray, and J. Midcap. 1986. Directory of Iowa nurseries that deal in forestry and ornamental planting stock. Bulletin Pm-956. Ames: Iowa State University Cooperative Extension Service.

Iowa Department of Natural Resources. 1988. *Prairie Establishment and Management Guide.* Des Moines: Office of County Conservation Activities, Iowa Department of Natural Resources.

Isaacson, R. T. 1989. *Andersen Horticultural Library's Source List of Plants and Seeds.* St. Paul: Andersen Horticultural Library, University of Minnesota.

Soil and Water Conservation Society. 1987. *Sources of Native Seeds and Plants.* Ankeny, Iowa.

Whealy, K. 1989. *Fruit, Berry, and Nut Inventory.* Decorah, Iowa: Seed Saver Publications.

Wray, P. H. 1991. Forestry contacts and organizations in Iowa. Forestry Extension Notes F-340. Ames: Iowa State University Cooperative Extension Service.

Index

Illustrations are indicated by italicized page numbers.

Alfisols, 59
American bittern, 91
American coot, 91
American elm, 47, 49, 50, 88
American hazelnut, 47
American hornbeam, 49
American kestrel, 23
American redstart, 56
American toad, 23
American woodcock, 55
amphibians, 92
Anderson Prairie, *11*
aquatic invertebrates, 92
aromatic aster, 19
arrowheads, 81
artificial drawdown. *See* wetland
   maintenance

badger, 22
bald eagle, 90, 91
balsam fir, 52
barnyard grass, 81
barred owl, 55
bass, 92
basswood, 48
beaver, 92
bedrock, 2, 3
bedstraw, 48
beggar-ticks, 52, 81
Bell's vireo, 56
benefits of ecosystem restoration,
   5; of forest restoration, 54–57; of

prairie restoration, 19–23; of
   wetland restoration, 88–93
big bluestem, 16, 18
bigtooth aspen, 46, 52
biological diversity, 6
Bishop's cap, 50
bison, 8, 24
bitternut hickory, 49, 50
black-and-white warbler, 56
black ash, 49, 50
black bear, 57
blackberry, 49
black-capped chickadee, 55
black cherry, 46, 49
black-crowned night heron, 91
blackjack oak, 46
black locust, 76
black maple, 48, 49
black oak, 46
black snake-root, 53
black walnut, 49, 50
black willow, 50, 51
bladder fern, 53
bladdernut, 49
blazing star, 19
bloodroot, 50
blueflag iris, 15
blue grama grass, 19
blue jay, 55
bluejoint reedgrass, 15
blue-winged teal, 90
bobolink, 22

bottle gentian, 15
bottomland hardwoods community.
    See forests
box elder, 51, 76
bristly buttercup, 50
broad-winged hawk, 56
brown-headed cowbird, 23
bulrushes, 81, 83
bur oak, 46, 47, 48
bur reeds, 83
buttercup family, 10
butternut, 49, 50
buttonbush, 50

Canada goldenrod, 17
Canada goose, 90
Canada yew, 52
catfish, 92
cattails, 81, 83, *83*, 87, 100, *101*
channelization. *See* wetlands
chemical weed control. *See* herbi-
    cide use
chinkapin oak, 46, 53
chokecherry, 47, 50
Clear Lake, 87
climate: current, 5; historical cli-
    mate change, 4, 8, 42
columbine, 53
commercial nurseries, 105–113;
    forest plants, 107–112; prairie
    plants, 105–106; wetland plants,
    113
common loon, 90
common mountain mint, 15
common reed, 81
common yellowthroat, 23, 91
concentric plant community zones.
    See wetlands
coontails, 83

cost: of forest restoration, 77; of
    prairie restoration, 40–41; of
    wetland restoration, 101–102
cost sharing: for forest establish-
    ment, 77; for wetland establish-
    ment, 98
cottonwood, 50, 51, 88
cover crop for prairie establishment,
    33
cowbane, 15
crownbeard, 54

daisy (composite) family, 10
deer mouse, 56
degenerating marsh. *See* palustrine
    wetland dynamics
Des Moines Lobe, *1*, 3, 15
dickcissel, 22
dogwoods, 48, 49
Dolliver Memorial State Park, *52*
Doolittle Prairie, *15*, *17*, *82*
drought, 8
dry marsh. *See* palustrine wetland
    dynamics
dry-mesic prairie community. *See*
    prairies
dry prairie community. *See* prairies
duckweeds, 83
dutchman's breeches, 50

eared grebe, 90
eastern bluebird, 55
eastern chipmunk, 56
eastern cottontail, 56
eastern kingbird, 23
eastern red cedar, 53
eastern wood-pewee, 55
economic assets, 6
elderberry, 50, 51

elk, 22

emergent community. *See* wetlands

erect dayflower, 19

ermine, 56

Euro-American settlement: conversion of forests, 43–44; conversion of prairies, 8–9; conversion of wetlands, 78–79; effects on landscape ecology, 1

evening bat, 56

evening primrose, 19

extirpated wildlife species: of forests, 56–57; of prairies, 22; of wetlands, 90

fall panicum, 81

false spikenard, 50

false sunflower, 17, 53

fameflower, 19

fens, 84–85

Field Extension Education Laboratory, *30, 31*

fire: historical role of fire, 7–8; in savannas, 48; use in prairie management, 37–40; use in wetland management, 100

firebreaks: for forest protection, 59; for prairie maintenance, 25, 39

fishes, 92

fleabane, 19

floating-leaved community. *See* wetlands

flowering spurge, 53

flycatchers, 55

forbs, 9

forest community composition. *See* forests

forest establishment, 57–74; planting methods, 68–74; site preparation, 68; site selection, 58; sources of propagules, 74; species to plant, 60

forest interior wildlife species, 56, 59

forest maintenance, 74–77; specific forest products, 77; weed control in, 75–76

forests, 1, 5, 42–57; benefits of forest areas, 54–57; forest communities, 45–54; history of, 42–44

Forster's tern, 91

fox squirrel, 56

fragile fern, 50

Franklin's ground squirrel, 22

free-floating community. *See* wetlands

frogs, 92

game-bird species, 91

genetic resources of natural plant communities: forests, 44; prairies, 19–20

geology and landforms, *1*, 2–3

giant bur reed, 81

ginseng, 53

glaciation, 2, 42, 78

glyphosate. *See* herbicide use

golden alexanders, 17

goldenrods, 48, 51, 53

grapes, 48, 50

grasses, 9–10, 81

grasshopper mouse, 22

grasshopper sparrow, 22

grass-of-Parnassus, 85

gravel hill prairies, 19

gray dogwood, 48

gray fox, 56

grayhead prairie coneflower, 16

gray squirrel, 56
gray wolf, 57
great blue heron, 91
great egret, 91
greater prairie-chicken, 22
great-flowered beardtongue, 18
great horned owl, 55, *56*
green ash, 50, 88
green-backed heron, 55, 91
green-winged teal, 90

hackberry, 47, 50, 88
hairy grama grass, 19
Hayden, Ada, 19
Hayden Prairie, 9, *16*
heath aster, 17, 18
hemi-marsh stage. *See* palustrine
    wetland dynamics
Hendrickson Marsh, *91*
Henslow's sparrow, 22
hepatica, 49
herbicide use: for forest site prepa-
    ration, 68; for maintenance of
    tree seedling plantings, 76, 101;
    for prairie site preparation, 28–29
highbush cranberry, 52
high-grading, 43. *See also* history of
    forests
history of forests, 42–44; forest de-
    cline, 43–44; presettlement for-
    est area, 43
history of prairies, 7–9; loss of prai-
    ries, 9; presettlement prairie area,
    7–8
history of wetlands, 78–80; loss of
    wetlands, 79–80; presettlement
    wetland area, 79
hog peanut, 48
honey locust, 50

hooded ladies' tresses, 84
hooded merganser, 55
hops, 53
hydric soils, 93–95
hydrophytic plants, 80

impatiens, 51
Indiana bat, 56
Indiangrass, 17, 18
indigo bush, 50
inoculation of seed (prairie le-
    gumes), 31–32
Iowan Surface, *1*, 3
Iowa state prairie preserves, 9
ironweed, 54
ironwood, 47, 49

jack-in-the-pulpit, 50

Keen's myotis, 56
Kentucky coffeetree, 50
kidneyleaf buttercup, 50
kinglets, 55
king rail, 91
knotgrass, 19

lacustrine wetlands, 87
lake marsh. *See* palustrine wetland
    dynamics
large arrow grass, 84
leadplant, 17, 18
least shrew, 56
least tern, 90
least weasel, 22, 56
leatherwood, 53
legume family, 10
leopard frog, 23
little bluestem, 16, 18, 19
Little Wall Lake, 87

lizards, 92
local genotypes: for forest restoration, 60; for prairie restoration, 36; for wetland restoration, 98–99
loess, 3
Loess Hills prairies, 18–19
loggerhead shrike, 22
long-billed curlew, 22
long-eared owl, 56
long-tailed weasel, 22, 92
lynx, 57

Macbride, Thomas, 5
MacDonald, G. B., 43
mallard, 90
manna grass, 81
marbled godwit, 22
Marietta Sand Prairie, *18*
marsh muhly grass, 15
masked shrew, 22
Mason State Tree Nursery, *35*
meadow bedstraw, 17
meadow jumping mouse, 22
meadowlarks, 22
meadow vole, 22
mesic prairie community. *See* prairies
mice, 92
milkweed family, 10
mink, 92
minnow, 92
mints, 10, 65
Mississippi River, 2, 79, 80
Missouri River, 2, 79
model ecosystems: model forest communities, 58; prairie as model for agriculture, 21
Mollisols, 24

moonseed, 53
mountain maple, 52
mowing: for forest maintenance, 76; for prairie establishment, 37; for wetland maintenance, 100
mudflat annual community. *See* wetlands
muskrat, 86, 92, 96

nannyberry, 47
needlegrass, 18, 19
New England aster, 15
northern bog orchid, 85
northern bush honeysuckle, 53
northern conifer and hardwood community. *See* forests
northern harrier, 22
northern oriole, 55
northern pin oak, 46, 47
northern pintail, 90
northern shoveler, 90
Northwest Iowa Plains, *1*, 3
nuthatch, 55

oak-basswood community. *See* forests
oak-cedar limestone glade. *See* forests
oak-hickory forest community. *See* forests
oak savanna. *See* forests
old-field goldenrod, 19
opossum, 22, 56
otter, 92
ovenbird, 56

Paleozoic Plateau, *1*, 3
palustrine wetland dynamics, 85–86

palustrine wetlands, 81–86
Pammel, Louis, 5
paper birch, 52
parsley family, 10
pawpaw, 50
peachleaf willow, 50, 51
pecan, 50
perch, 92
pied-billed grebe, 91
pike, 92
pink milkwort, 19
piping plover, 90
plains garter snake, 23
plains muhly grass, 18, 19
plains pocket gopher, 22
plum, 51
poison ivy, 50
pondweeds, 83
porcupine, 57
post oak, 46
prairie cordgrass, 15, 81
prairie coreopsis, 18
prairie dropseed, 16, 18
prairie establishment, 23–37;
    planting methods, 29–35; site
    preparation, 27–29; site selec-
    tion, 24–27; sources of propa-
    gules, 36–37; species to plant, 27
prairie goldenrod, 18, 19
prairie maintenance, 37–40; mow-
    ing, 37; prescribed burning,
    37–40
prairie pothole region, 15, 78
prairie remnants, 9
prairie roadsides, 20, 22
prairies, 1, 5, 7–23; benefits of prai-
    rie areas, 19–23; history of, 7–9;
    prairie communities, 9–19
prairie skink, 22

prairie soils, 9, 19, 24
prairie vole, 16
prairie-wetland complexes, 1, 15
prairie wild rose, 18
prescribed burning. See fire
prickly ash, 47
prickly gooseberry, 47
purple meadow rue, 15
purple milkwort, 19
pye-weeds, 53

quaking aspen, 46, 52

raccoon, 23, 56, 92
ragweed, 81
red-backed vole, 57
red baneberry, 52
red bat, 56
red elderberry, 52
red fox, 22, 92
redhead, 90
red mulberry, 47
red oak, 47, 48
red osier dogwood, 48, 52
red-shouldered hawk, 56, 91
red squirrel, 56
red-tailed hawk, 55
red three-awn, 18
red-winged blackbird, 91
reed canary grass, 81
reeds, 83
regenerating marsh. See palustrine
    wetland dynamics
reptiles and amphibians, 92
ring-necked pheasant, 23
riparian forest community. See
    forests
riparian wetlands, 87–88

river birch, 50, 51
riverine wetlands, 88
rock elm, 50
rose-breasted grosbeak, 55
rose family, 10
rough blazing star, 19
ruby-throated hummingbird, 55
ruddy duck, 90
rufous-sided towhee, 56

salamanders, 92
sandbar willow, 51
sand dropseed, 18, 19
sandhill crane, 90
sand lily, 18
sand prairies. *See* prairies
sassafras, 53
savanna, 1, 48
sawtooth sunflower, 16
scarification, 30, 31
scarlet guara, 18
scarlet tanager, 56
Scribner's panic grass, 17, 18
sedges, 15, 16, 48, 50, 52, 81
sedge wren, 22
seed dealers: prairie seed, 36,
    105–106; wetland seed, 113
seed drills, 34
seeding rates for prairie establish-
    ment, 32–33
seepage wetlands. *See* fens
serviceberry, 47, 49, 53
shagbark hickory, 46, 47
sharp-shinned hawk, 56
shellbark hickory, 50
Shimek, Bohumil, 5
shiner, 92
shingle oak, 46, 50
short-eared owl, 22

short-tailed shrew, 56
showy sunflower, 17
shrews, 92
shrubs: in forests, 45–53; planting
    in forest restoration, 74; in prai-
    ries, 9
sideoats grama grass, 18
Silver Lake Fen, 85
silver maple, 50, 51, 88
skeleton weed, 18
skippers, 23
Skunk River greenbelt, *47*
sky-blue aster, 17
slender rockbrake fern, 53
slippery elm, 47, 88
small fringed gentian, 85
small white lady's slipper, 85
smartweed, 52, 81
smooth sumac, 48
snakes, 92
snow bunting, 23
snow goose, 90
soils, 3–4; forest soils, 59; prairie
    soils, 9, 19, 24; wetland soils,
    93–96
Solomon's seal, 48
song sparrow, 23
southern flying squirrel, 56
Southern Iowa Drift Plain, *1*, 3
speckled alder, 52
spikenard, 53
spike rushes, 81
Spirit Lake, 87
spotted sandpiper, 91
spotted skunk, 22
starry campion, 53
sticklebacks, 92
stiffstem flax, 19
Stinson Prairie, 7

stratification: of forest seed, 69; of prairie seed, 30–31
striped skunk, 22, 92
submersed community. *See* wetlands
succession in forests, 57–58
sugar maple, 48, 49
sunfish, 92
swallow-tailed kite, 90
swamp saxifrage, 15
swamp white oak, 50
sycamore, 50

Tazewell Drift, *1*, 3
thirteen-lined ground squirrel, 22
toads, 92
toothwort, 50
transplanting techniques: for forest seedlings, 71–74; for prairie establishment, 34–35
tree-planting machines, 71, *72*
tree swallow, 55
trout, 92
trout lily, 50
trumpeter swan, 90
tufted titmouse, 55
turkey vulture, 55
turtles, 92
Type I–Type IV wetlands, 81–83

upland sandpiper, 22

vervain, 51
vesper sparrow, 23
vireos, 55, 91
Virginia creeper, 48, 50
Virginia waterleaf, 51
voles, 92

wahoo, 49, 50
warblers, 55, 91
waterlilies, 83
watermilfoils, 83
weed control. *See* forest establishment; prairie establishment
western harvest mouse, 22
wetland establishment, 93–99; establishing vegetation, 98–99; site preparation, 96–98; site selection, 93–96
wetland maintenance, 99–101
wetlands, 1, 5, 78–93; benefits of wetland areas, 88–93; history of, 78–80; wetland communities, 80–88
wetland seed bank, 96
wet meadow community. *See* wetlands
wet-mesic prairie community. *See* prairies
wet prairie community. *See* prairies
whippoorwill, 55
white ash, 46, 49
white-footed mouse, 56
white oak, 46, 47, 48, 49
white pine, 52
White Pine Hollow, *53*
white-tailed deer, 56, 92
white-tailed jackrabbit, 22
whooping crane, 90
wild lily-of-the-valley, 53
wild millet, 81
wild sarsaparilla, 50
wild strawberry, 16
wild turkey, 55
willows, 51, 88
winter wren, 56
witch hazel, 49

wolverine, 57
wood anemone, 48
woodchuck, 56
wood duck, 55, 90
woodland vole, 57
wood nettle, 51
woodpeckers, 55, 91
wood thrush, 56
worm-eating warbler, 56
wrens, 55

yellow birch, 52
yellow-crowned night heron, 55
yellow-headed blackbird, 91
yellow-lipped ladies' tresses, 84
yellow puccoon, 18
yellow stargrass, 15
yucca, 18

A Cook's Tour of Iowa
By Susan Puckett

The Folks
By Ruth Suckow

Fragile Giants: A Natural History of the Loess Hills
By Cornelia F. Mutel

An Iowa Album: A Photographic History, 1860–1920
By Mary Bennett

Iowa Birdlife
By Gladys Black

Landforms of Iowa
By Jean C. Prior

More han Ola og han Per
By Peter J. Rosendahl

Neighboring on the Air: Cooking with the KMA Radio Homemakers
By Evelyn Birkby

Nineteenth-Century Home Architecture of Iowa City: A Silver Edition
By Margaret N. Keyes

Nothing to Do but Stay: My Pioneer Mother
By Carrie Young

Old Capitol: Portrait of an Iowa Landmark
By Margaret N. Keyes

Parsnips in the Snow: Talks with Midwestern Gardeners
By Jane Anne Staw and Mary Swander

A Place of Sense: Essays in Search of the Midwest
Edited by Michael Martone

Prairies, Forests, and Wetlands: The Restoration of Natural Landscape Communities in Iowa
By Janette R. Thompson

A Ruth Suckow Omnibus
By Ruth Suckow

"A Secret to Be Burried": The Diary and Life of Emily Hawley Gillespie, 1858–1888
By Judy Nolte Lensink

Tales of an Old Horsetrader: The First Hundred Years
By Leroy Judson Daniels

The Tattooed Countess
By Carl Van Vechten

"This State of Wonders": The Letters of an Iowa Frontier Family, 1858–1861
Edited by John Kent Folmar

Townships
Edited by Michael Martone

Vandemark's Folly
By Herbert Quick

The Wedding Dress: Stories from the Dakota Plains
By Carrie Young